Advanced Structured Materials

Volume 143

Common engineering materials reach in many applications their limits and new developments are required to fulfil increasing demands on engineering materials. The performance of materials can be increased by combining different materials to achieve better properties than a single constituent or by shaping the material or constituents in a specific structure. The interaction between material and structure may arise on different length scales, such as micro-, meso- or macroscale, and offers possible applications in quite diverse fields.

This book series addresses the fundamental relationship between materials and their structure on the overall properties (e.g. mechanical, thermal, chemical or magnetic etc.) and applications.

The topics of *Advanced Structured Materials* include but are not limited to

- classical fibre-reinforced composites (e.g. glass, carbon or Aramid reinforced plastics)
- metal matrix composites (MMCs)
- micro porous composites
- micro channel materials
- multilayered materials
- cellular materials (e.g., metallic or polymer foams, sponges, hollow sphere structures)
- porous materials
- truss structures
- nanocomposite materials
- biomaterials
- nanoporous metals
- concrete
- coated materials
- smart materials

Advanced Structured Materials is indexed in Google Scholar and Scopus.

More information about this series at http://www.springer.com/series/8611

Vladimir A. Polyanskiy · Alexander K. Belyaev
Editors

Advances in Hydrogen Embrittlement Study

Springer

Editors
Vladimir A. Polyanskiy
Institute for Problems in Mechanical Engineering
Russian Academy of Sciences
St. Petersburg, Russia

Alexander K. Belyaev
Institute for Problems in Mechanical Engineering
Russian Academy of Sciences
St. Petersburg, Russia

ISSN 1869-8433 ISSN 1869-8441 (electronic)
Advanced Structured Materials
ISBN 978-3-030-66947-8 ISBN 978-3-030-66948-5 (eBook)
https://doi.org/10.1007/978-3-030-66948-5

This Springer imprint is published by the registered company Springer Nature Switzerland AG
The registered company address is: Gewerbestrasse 11, 6330 Cham, Switzerland

Preface

Hydrogen greatly influences the properties of various materials which is important for almost all industrial technologies. Hydrogen can enter the molten metal and remain in it after solidification. During the exploitation of metal parts, hydrogen saturation occurs due to corrosion, friction, or plastic deformation. Water is the most common source of hydrogen. The properties of metals are drastically affected in the presence of any additives. They become brittle, crack, and lose durability. Only one hydrogen atom per hundreds of thousands of metal atoms is enough while other impurities can cause some harm only in much higher concentrations reaching a fraction of a percent. A small concentration of hydrogen is difficult to fix, which makes difficult the direct measurement of its content in metals. That is why various approaches to the problem of hydrogen diffusion and distribution in metals are attracting a vivid attention of the researchers from many countries.

In consideration of success of workshop on advances in hydrogen embrittlement study which took place in the framework of the First International Conference "Corrosion in the oil & gas industry" in St. Petersburg in May 2019, we feel honoured and privileged to present a collection of papers in this volume "Advances in hydrogen embrittlement study" in the world leading publisher Springer. The workshop covered, among the others, the subjects that reflected some challenging research items in theoretical and experimental studies. These include both developing adequate experimental techniques for characterizing fracture/damage and efficient theoretical and numerical approaches. The volume contains predominantly the papers presented at the above workshop and completed by new scientific achievements in a number of papers by the world experts in the field of hydrogen–metal interaction.

St. Petersburg, Russia
November 2020

Vladimir A. Polyanskiy
Alexander K. Belyaev

Contents

Chapter 1
Potential Effects of Short-Range Order on Hydrogen Embrittlement of Stable Austenitic Steels—A Review

Motomichi Koyama, Burak Bal, Dermican Canadinc, Kishan Habib, Toshihiro Tsuchiyama, Kaneaki Tsuzaki, and Eiji Akiyama

Abstract Here, we present a review of the hydrogen embrittlement behavior of face-centered cubic (FCC) alloys with short-range order (SRO) of solute atoms. In this paper, three types of FCC alloys are introduced: Fe–Mn–C austenitic steels, high-nitrogen steels, and CoCrFeMnNi high-entropy alloys. The Fe–Mn–C austenitic steels show dynamic strain aging associated with Mn–C SRO, which causes deformation localization and acceleration of premature fracture even without hydrogen effects. The disadvantageous effect of dynamic strain aging on ductility, which is associated with the deformation localization, amplify plasticity-assisted hydrogen embrittlement. Cr–N and Co–Cr–Ni SRO effects in high-nitrogen austenitic steels and high-entropy alloys enhance the dislocation planarity, which causes stress concentration in the grain interior and near the grain boundaries. The stress concentration coupled with hydrogen effects causes quasi-cleavage and intergranular fractures.

Keywords Hydrogen embrittlement · Austenitic steel · Short-range order · Dynamic strain aging · Dislocation planarity

---

M. Koyama (✉) · E. Akiyama
Institute for Materials Research, Tohoku University, Katahira 2–1–1, Aoba–ku, 980-8577 Sendai, Miyagi, Japan
e-mail: koyama@imr.tohoku.ac.jp

M. Koyama · K. Tsuzaki
Elements Strategy Initiative for Structural Materials (ESISM), Kyoto University, Yoshida-honmachi, Sakyo–ku, 606–8501 Kyoto, Japan

B. Bal
Department of Mechanical Engineering, Abdullah Gül University, 38080 Kayseri, Turkey

D. Canadinc
Department of Mechanical Engineering, Koc University, 34450 Sariyer, Istanbul, Turkey

K. Habib · K. Tsuzaki
Research Center for Structural Materials, National Institute for Materials Science, 1–2–1 Sengen, 305–0047 Tsukuba, Japan

T. Tsuchiyama · K. Tsuzaki
Faculty of Engineering, Kyushu University, Motooka 744, Nishi-ku, 819-0395 Fukuoka, Japan

V. A. Polyanskiy and A. K. Belyaev (eds.), *Advances in Hydrogen Embrittlement Study*, Advanced Structured Materials 143, https://doi.org/10.1007/978-3-030-66948-5_1

1.1 Introduction

Austenitic steels have been noted for their hydrogen-resistance, owing to the low hydrogen diffusivity and isotropic plasticity levels stemming from their face-centered cubic (FCC) structure. The former prevents hydrogen localization at dislocations and microstructural interface compared with body-centered cubic (BCC) structure, and the latter enables accommodation of the microstructural stress concentration under loading because of the larger number of slip systems compared with hexagonal close-packed (HCP) structure. In this context, even if the initial crystallographic structure is FCC, a martensitic transformation to a BCC [1–4] or HCP [5–8] structure triggers the occurrence of hydrogen embrittlement. Therefore, with the scope of application in hydrogen-related infrastructures, austenite has been stabilized through the addition of solute elements such as Ni, Cr, Mn, C, and N. Figure 1.1 shows the hydrogen embrittlement susceptibility plotted against the equivalent Ni concentration, Ni_{eq} (austenite phase stability increases with Ni_{eq}.), which clearly indicates that the increase in austenite stability improves hydrogen embrittlement resistance when Ni_{eq} is below 40%.

However, we additionally note that there are some exceptions showing significant hydrogen embrittlement even when the austenite phase is fully stabilized. For instance, hydrogen embrittlement susceptibility deteriorated when Ni_{eq} increased by over 40%, as shown in Fig. 1.1. Furthermore, stacking fault energy has been proposed as another criterion for screening hydrogen embrittlement resistance [9, 10].

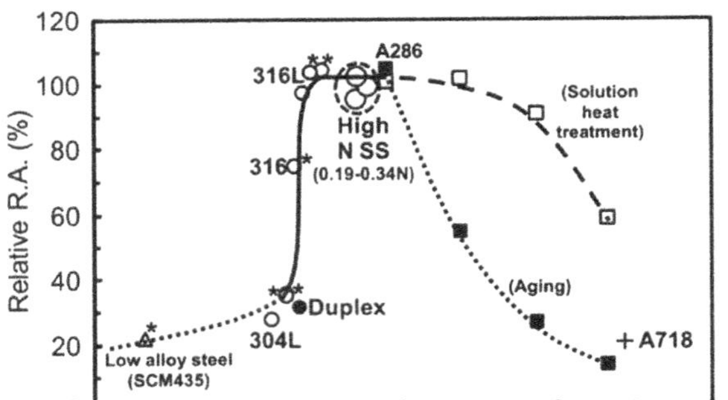

Fig. 1.1 Austenite stability effect on hydrogen embrittlement susceptibility when hydrogen was introduced at a hydrogen gas pressure of 45 MPa. Ni_{eq} indicates the austenite phase stability [2]. Specifically, the effects of various solute elements on the austenite stability were converted to the effect of Ni in terms of the thermodynamically equivalent Ni concentration in mass %. A more quantitative relationship between Ni_{eq} and solute element concentration is shown in the label of the X-axis. 304L, 316, and 316L indicate commercial austenitic stainless steels. A286 and A718 indicate precipitation-strengthened austenitic stainless steel and nickel alloy, respectively. Relative R.A. indicate relative reduction in area. Reproduced with permission from *ISIJ International*, **56**, 405–412 (2016) Copyright 2016, ISIJ

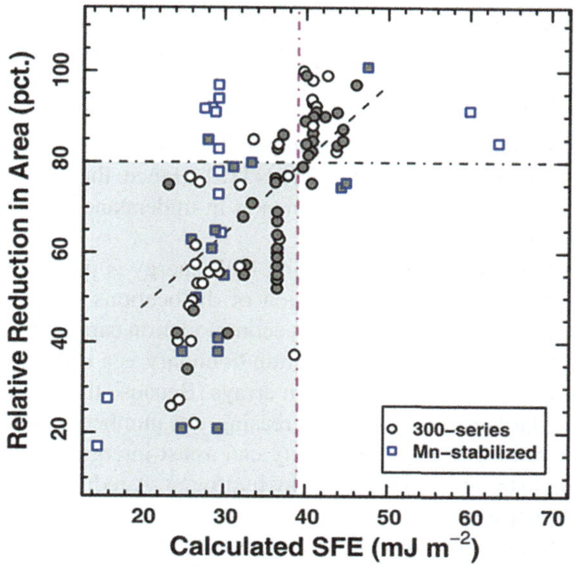

Fig. 1.2 Stacking fault energy criterion for screening hydrogen-resistant ferrous alloys [9]. 300-series and Mn-stabilized indicate austenitic stainless steels and high-Mn steels, respectively. The stacking fault energies were calculated for room temperature. The third dashed line is a fit to all the data in the range of 25 to 50 mJ/m^2 stacking fault energy. Shaded and open points indicate specimens tested after being saturated with internal hydrogen and specimens tested in external hydrogen, respectively. Reproduced with permission from *JOM*, **72**, 1982–1992 (2020) Copyright 2020. This is a U.S. government work and not under copyright protection in the U.S.

As shown in Fig. 1.2, remarkable hydrogen embrittlement occurred when stacking fault energy was lower than 39 mJ/m^2, regardless of whether martensitic transformation occurred. Therefore, we must recognize that the austenite stability criteria to predict the hydrogen embrittlement resistance of austenitic steels are incomplete, and that further significant factors triggering the hydrogen embrittlement of stable austenitic steels exist.

In general, the factors affecting embrittlement are local stress/plastic strain and the strength of microstructural interfaces, such as grain boundaries. To further specify from a metallurgical aspect, local stress/plastic strain are factors associated with the dislocation density evolution behavior and dislocation planarity. The dislocation density evolution is dependent on various factors related to recovery and dislocation multiplication. In terms of dislocation multiplication, dislocation pinning by solute atoms has a significant effect. Specifically, dynamic strain aging, which occurs during plastic deformation, provides high dislocation density and is associated with extraordinary work hardening [11, 12]. The high work-hardening capability delays the satisfaction of the plastic instability condition, which increases elongation. The occurrence of hydrogen embrittlement may be assisted by an increase in dislocation density; however, when local dislocation density has a detrimental effect on local stress/plastic strain at cracking sites. In fact, when austenitic steels contain Mn and C simultaneously, distinct dynamic strain aging occurs [13, 14]. Dynamic strain aging affects hydrogen embrittlement properties by changing local plastic deformation behavior [15, 16]. Therefore, the hydrogen embrittlement behavior of Fe–Mn–C austenitic steels is relatively atypical compared to that of conventional austenitic stainless steels [17] and binary Fe–Mn alloys [18]. Hence, the effect of Mn–C SRO on dynamic strain aging is a key phenomenon in understanding some exceptional behaviors of hydrogen embrittlement.

Regarding dislocation planarity, stacking fault energy is the primary controlling factor in FCC alloys because the extension of dislocations suppresses cross slip. The suppression of cross slip reduces the accommodation capability at the obstacles blocking dislocations. For instance, the grain boundary is a representative obstacle that stops the motion of planar dislocation arrays. Because the stress concentration at the grain boundary increases with increasing the number of piling-up dislocations, enhancement of dislocation planarity can assist intergranular cracking when the grain boundary strength is weakened by hydrogen. A reduction in stacking fault energy causes planar dislocation motion in addition to the promotion of twinning and martensitic transformation, which increases the hydrogen embrittlement susceptibility (Fig. 1.2). However, even when the stacking fault energy is not very low, high-nitrogen austenitic steels have shown strong dislocation planarity [19, 20] and according hydrogen embrittlement [21, 22]. Specifically, the dislocation planarity of the high-nitrogen steels have been discussed to arise from Cr–N coupling/clustering, *i.e.*, short-range ordering (SRO) [23, 24]. Therefore, atomic interactions such as Mn–C and Cr–N can be significant factors that control the hydrogen embrittlement susceptibility of stable austenitic steels.

Evidently, significant amounts of varying solute elements must be added to stabilize austenite and increase its strength, *for example*, Fe–Mn–C twining-induced plasticity steel and high-nitrogen steel. Hence, many advanced austenitic steels involve the effects of SRO. Therefore, understanding the short-range order effects is of crucial importance to fine-tune the alloy design strategy for simultaneous improvements in strength and ductility, both in air and hydrogen environments. In this paper, we present a focused review of our recent work on SRO-related hydrogen embrittlement in stable austenitic steels.

1.2 Mn–C Interaction

1.2.1 Case: High-Stacking Fault Energy

Initially, we introduced the effect of Mn–C on plastic deformation and hydrogen embrittlement. The Mn–C interaction has been reported to affect the behavior of dynamic strain aging, which results in serrated flow and enhances work harden-

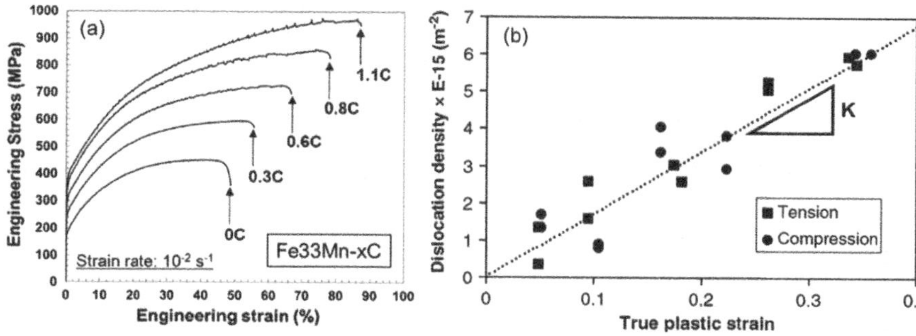

Fig. 1.3 **a** Engineering stress–strain curves of Fe–33Mn–xC austenitic steels (mass %), which show serrated flow associated with dynamic strain aging [12] **b** dislocation density increment with plastic strain in a Fe–14Mn–1C steel (mass %) [33]. Reproduced with permission from *Scripta Materialia*, **141**, 20–23 (2017) and **55**, 299–302 (2006) Copyright 2016, Elsevier

ing [25, 26].[1] Here, we introduce the effect of Mn–C-related dynamic strain aging on hydrogen embrittlement in the case of a relatively high-stacking fault energy in a group of high-Mn steels (over 50 mJ/m^2), because the deformation twinning formed by reductions in stacking fault energy complicates the hydrogen embrittlement behavior.

The effect of dynamic strain aging in Fe–Mn–C austenitic steels becomes distinct with increasing solute carbon content, as shown in Fig. 1.3a. Since most high-performance high-Mn steels contain more than 0.6 wt. % C, the effect of dynamic strain aging on plastic deformation behavior is always significant. Specifically, the dynamic strain aging facilitates pinning of dislocations at an increased rate, resulting in dislocation multiplication for further plastic deformation (Fig. 1.3b). Therefore, the dislocation accumulation rate in Fe–Mn–C austenitic steels is markedly higher than in other FCC materials, as listed in Table 1.1. Furthermore, dynamic strain aging triggers deformation localization [28–30]. The localized extraordinary accumulation of dislocations deteriorates the ductility associated with deformation band cracking [31] and transgranular void formation [12] even without hydrogen uptake. Hydrogen accelerates the reduction in ductility (Fig. 1.4a), which results from transgranular and intergranular fractures (Fig. 1.4b). Note here that the intergranular fracture surface involves traces of plastic deformation, as shown in the inset of Fig. 1.4c. Furthermore, the morphology of the transgranular cracks is wavy, indicating that the cracked surface does not have any specific crystallographic features and is not along deformation bands such as twin plates [32]. The dislocation microstructure between the deformation twins is either a cell or a dislocation wall (Fig. 1.4d). These facts imply that a transgranular crack is formed by dislocation accumulation and interactions.

Intrinsically, hydrogen embrittlement becomes significant with decreasing strain rate, because it gives a longer time for hydrogen diffusion, accumulation, hydrogen-

[1]Currently, the effect of Mn–C interaction is still under debate. A recent work pointed out a type of dynamic strain aging is triggered by a pipe diffusion of carbon without Mn–C SRO [27].

Table 1.1 Increasing rate of dislocation density per unit strain (K) in various FCC alloys. The Fe–Mn–C Hadfield steel shows a significantly higher dislocation accumulation rate than the others [25, 33]

Material	Deformation	K (m^{-2} per unit strain)	References
Aluminum	Torsion	2.9×10^{15}	[34]
Copper	Torsion	3.1×10^{15}	[34]
Copper	Compression	1.5×10^{15}	[35]
Nickel	Torsion	2.0×10^{15}	[36]
γ phase of type 304 austenitic stainless steel	Rolling	1.3×10^{15}	[37]
Fe-Mn-Si-Al TWIP steel	Tension	4.0×10^{15}	[38]
Fe-14Mn-1C steel (Hadfield steel)	Tension or compression	1.7×10^{16}	[33]

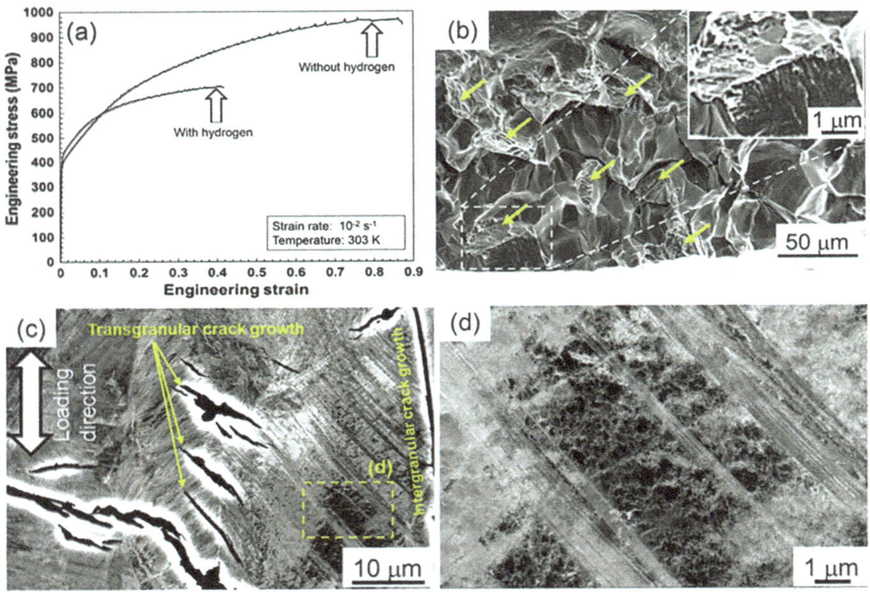

Fig. 1.4 **a** Engineering stress–strain curves of the Fe–33Mn–1.1C austenitic steel (mass%) with and without hydrogen precharging. Hydrogen charging was performed at a current density of 7 A/m^2 at 353 K for 72 h in a 3% NaCl aqueous solution containing 3 g/L NH_4SCN. **b** Fracture surface and **c** transgranular sub-cracks of the fractured specimen with hydrogen precharging. **d** Magnification of **c**. [39] Reproduced with permission from *ISIJ International*, **56**, 405–412 (2016) Copyright 2016, Elsevier

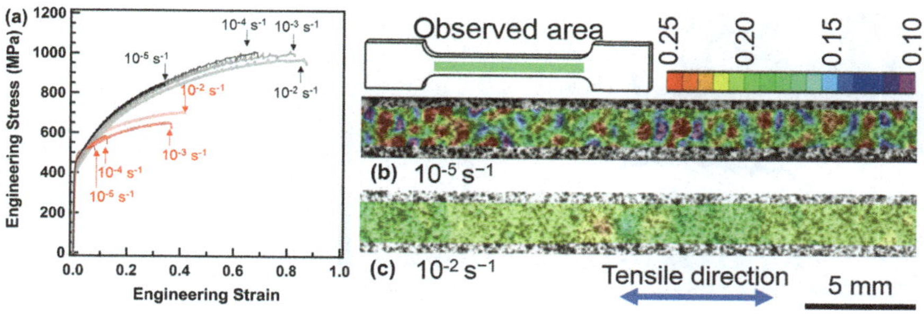

Fig. 1.5 **a** Strain rate effect on mechanical degradation in the Fe–33Mn–1.1C steel (mass%) with and without hydrogen precharging. Hydrogen charging was performed at a current density of 7 A/m^{2}at 353 K for 72 h in a 3% NaCl aqueous solution containing 3g/L NH_4SCN. **b, c** Strain contour maps of the uncharged specimen deformed to 20% macroscopic tensile strain at initial strain rates of 10^{-5} and 10^{-2} s^{-1}. [40] Reproduced with permission from *Metallurgical and Materials Transactions A*, **50**, 1137–1141 (2019) Copyright 2019, The minerals, Metals & Materials Society and ASM International

dislocation interaction. Moreover, decreasing the strain rate decreases the dislocation velocity, which increases the interaction time between dislocation and carbon (or Mn–C). Therefore, the strain rate effect of ductility degradation in Fe—Mn–C austenitic steels with high-stacking fault energy is extremely remarkable, as shown in Fig. 1.5a. The strain rate effect is associated with the multi-scale phenomenon. That is, the appearance of mesoscopically localized deformation bands, which is called Portevin Le-Chatelier (PLC) band, becomes clear as the strain rate decreases, even without hydrogen (Fig. 1.5b, c). In the PLC band, hydrogen reacts with dislocations and grain boundaries, which assists the occurrence of embrittlement under the low strain rate regime.

1.2.2 Case: Low Stacking Fault Energy

When the stacking fault energy is reduced to 20–40 mJ/m^2 in high-Mn austenitic steels, deformation twinning acts as a primary work-hardening mechanism. These steels are called twinning-induced plasticity (TWIP) steels [41, 42]. Twinning-assisted hardening can homogenize plastic strain distribution, which thereby suppresses the effects of dynamic strain aging on ductility degradation. For instance, Fe–Mn–C TWIP steels without hydrogen charging do not show distinct degradation of uniform elongation with decreasing strain rate [43]. However, the occurrence of dynamic strain aging is still important to consider the hydrogen embrittlement process. It should be noted that a portion of hydrogen-induced cracks propagates along twin boundaries [44–46] (Fig. 1.6a). The factor causing the twin boundary cracking is a localization of hydrogen in the deformation twins [46, 47]. According to an experimental study, hydrogen can be trapped at twin boundaries [48]. However, it

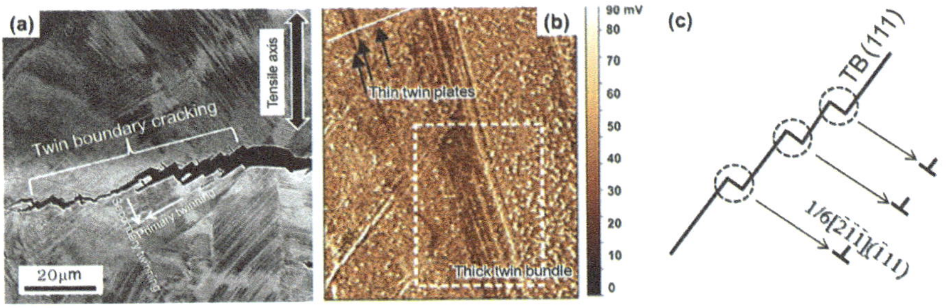

Fig. 1.6 **a** Cracking along deformation twins in a hydrogen-charged Fe–18Mn–1.2C steel [45]. **b** Surface potential figure showing hydrogen localization in the deformation twin plates, which was obtained by scanning Kelvin probe force microscopy [47, 54]. **c** Schematic showing the formation of steps on a coherent twin boundary via dislocation dissociation. [45] Reproduced with permission from *Acta Materialia*, **61**, 4607–4618 (2013) Copyright 2013, Elsevier, and *Journal of Electrochemical Society*, **160**, C643 (2015). Copyright 2015, The Electrochemical Society

has been theoretically clarified that hydrogen cannot be trapped in a coherent twin boundary [49]. To enable hydrogen entrapment at twin boundaries, the boundary character must be incoherent. It is known that dislocation reactions at coherent twins can form steps on the boundary [50, 51], which creates an incoherent portion [52]. It is considered that the incoherent portion of twin boundaries would act as a trap site for hydrogen [45, 53]. In this context, the high accumulation rate of dislocations per strain can increase the frequency of dislocation–twin boundary interactions, thus creating an incoherent portion on the twin boundaries, which results in twin boundary cracking.

Furthermore, the effect of strain aging also affects the delayed fracture property. Figure 1.7a shows the engineering stress–strain curves with a displacement holding process for 10 h at 69% strain. The 69% deformation strain was performed in air, and hydrogen charging was performed during the displacement holding. Interestingly, the time period until fracture during the displacement holding decreased with decreasing pre-strain rate (Fig. 1.7b). A cause of the pre-strain rate effect was that the flow stress increased with decreasing pre-strain rate. Because the work-hardening capability associated with dynamic strain aging increases by decreasing the strain rate, the external stress applied during the displacement holding increases as well, thus, accelerating the occurrence of delayed fracture. It should be noted that the delayed fracture was accelerated at the low pre-strain rate even after manipulating the pre-strain to make the external stress lower than that at the highest strain rate (Fig. 1.7c). This fact indicates that the effect of dynamic strain aging on the delayed fracture property results from both the variation in flow stress, as well as the microstructural changes (or deformation localization). Although the specific changes in microstructural evolution or deformation localization behavior are not clear, it can be concluded that Mn–C SRO associated dynamic strain aging is a disadvantageous factor in hydrogen embrittlement.

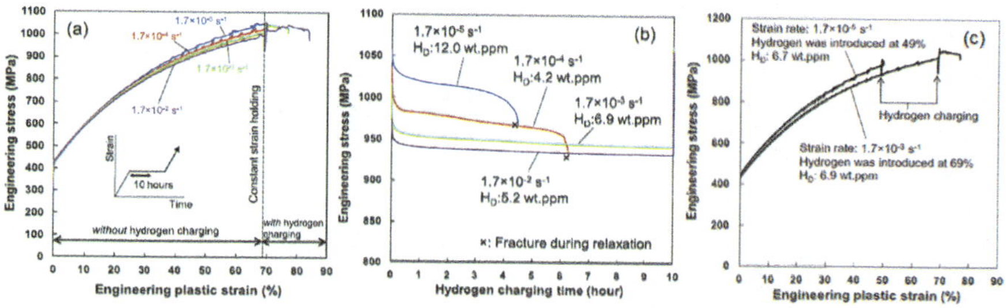

Fig. 1.7 **a** Engineering stress–strain curves including the process of displacement holding at the 69% strain with different pre-strain rates in Fe–22Mn–0.6C TWIP steel (mass %). Hydrogen charging was initiated immediately after the onset of the displacement holding and was performed at a current density of 7 A/m^2 and ambient temperature in a 3% NaCl aqueous solution containing 3g/L NH_4SCN. **b** Time for failure under displacement holding. **c** Delayed fracture behavior with different pre-strain and pre-strain rates. [15] Reproduced with permission from *Scripta Materialia*, **66**, 947–950 (2012). Copyright 2012, Elsevier

1.3 Cr–N Interaction

1.3.1 Grain Boundary Cracking

Following, we noted the effects of Cr–N interactions on the evolution of dislocation microstructures, which is associated with hydrogen embrittlement. Cr–N interactions have been noted in high-nitrogen austenitic steels. The formation of Cr–N couples results in planar dislocation slips (Fig. 1.8a, b), which increased the dislocation pile-up stress and were associated with the high work-hardening capability [55] (Fig. 1.8c). However, the stress concentration associated with the planar dislocation slips occasionally assisted brittle fracture even without hydrogen [19, 56, 57] (Fig. 1.8d).

Hydrogen charging causes remarkable degradation of ductility of high-nitrogen austenitic steels (Fig. 1.9a). The degradation of ductility results from the occurrence of intergranular and quasi-cleavage fractures. The intergranular fracture surface is shown in Fig. 1.9b. Note that the intergranular fracture surface contains slip marks and voids, which are traces of plastic deformation. Furthermore, the planar dislocation arrays and associated stress concentration, which corresponds to the schematic shown in Fig. 1.8c, are observed in the postmortem high-nitrogen steel specimen (Fig. 1.10a–c). This fact implies plasticity-driven intergranular cracking caused by Cr–N SRO via the planar dislocation slip, which is the primary cause of the hydrogen-induced degradation of ductility. In addition, the behavior of hydrogen-assisted cracking is dependent on the number of activated slip systems. When a single slip system is activated, the planar dislocation slip causes intergranular cracking. When multiple slip systems are activated, quasi-cleavage cracking occurs, as reported in the next section.

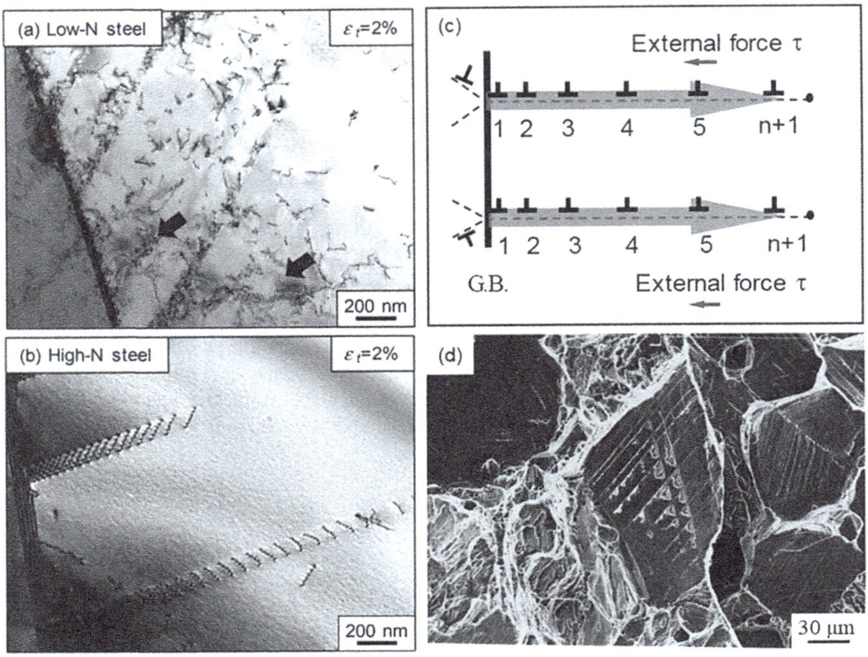

Fig. 1.8 Dislocation patterns of type 316L austenitic stainless steel (base chemical composition: Fe17.6Cr12.3Ni1.19Mn0.55Si2.08Mo0.012C0.04N in mass %) **a** without extra nitrogen and **b** with 0.45 mass % nitrogen in solute [55]. **c** Schematic depicting the occurrence of stress concentration stemming from planar dislocation slip [55]. **d** Occurrence of quasi-cleavage fracture at low temperature in a Fe–19Mn–17Cr-0.5N austenitic steel without hydrogen [19]. Reproduced with permission from *Materials Transactions*, **61**, 678–684 (2020). Copyright 2020, JIM, and *Acta Materialia*, **46**, 1577–1587 (1998) Copyright 1998, Elsevier

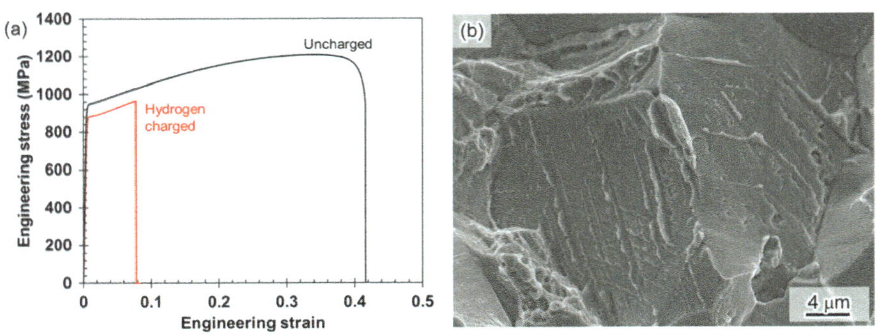

Fig. 1.9 **a** Engineering stress–strain curves of the Fe–25Cr–1N steel (mass%) with and without hydrogen pre-charging. Hydrogen was introduced at a hydrogen gas pressure of 100 MPa. **b** Presence of slip marks and voids on the intergranular fracture surface of the hydrogen-charged steel [21]. Reproduced with permission from *International Journal of Hydrogen Energy*, **45**, 10209-10218 (2020). Copyright 2020, Elsevier

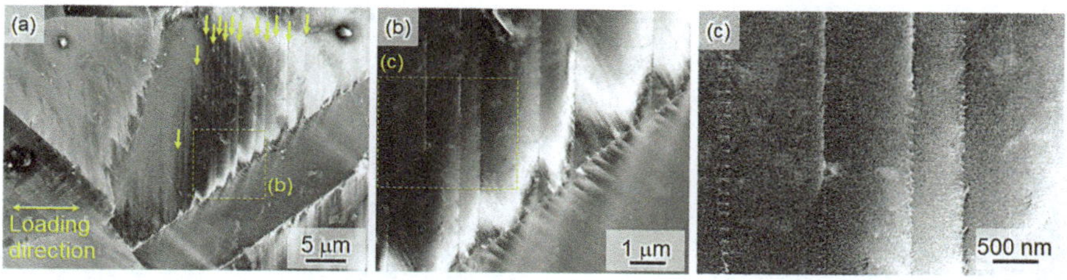

Fig. 1.10 Dislocation planar arrays with different observation scales in the hydrogen-charged high-nitrogen steel after fracture, which correspond to Fig. 1.9 [21]. The dashed squares in **a**, **b** indicate the observation regions of (**b**) and (**c**), respectively. Reproduced with permission from *International Journal of Hydrogen Energy*, **45**, 10209–10218 (2020). Copyright 2020, Elsevier

1.3.2 Transgranular Cracking

When multiple slip systems were activated, hydrogen-assisted quasi-cleavage cracking occurred. Numerous slip marks appeared on the quasi-cleavage fracture surface of the high-nitrogen steel, indicating the occurrence of dislocation-driven cracking Fig. 1.11. The microscopic stress distribution was qualitatively visualized as the ECC contrast gradient to evaluate the cause of the quasi-cleavage cracking. The ECC contrast gradient near the grain boundaries in multi-slip grains was less than that in the single-slip grain. Instead, the multiple slip of extended dislocations results in a Lomer–Cottrell lock, which creates sessile dislocations. The sessile dislocations act as strong obstacles against dislocation motion [58], resulting in stress concentration in the grain interior (Fig. 1.12a). Accordingly, quasi-cleavage cracking occurs along the slip plane (Fig. 1.12b). Hence, the Cr–N driven planar dislocation motion causes a distinct tensile orientation dependence of the cracking behavior, which results in a mixed fracture mode consisting of intergranular and quasi-cleavage fracture features.

Fig. 1.11 Quasi-cleavage fracture surface of the hydrogen-charged Fe–25Cr–1N steel, which corresponds to (Fig. 1.9 and 1.10 [21]. Reproduced with permission from *International Journal of Hydrogen Energy*, **45**, 10209–10218 (2020). Copyright 2020, Elsevier

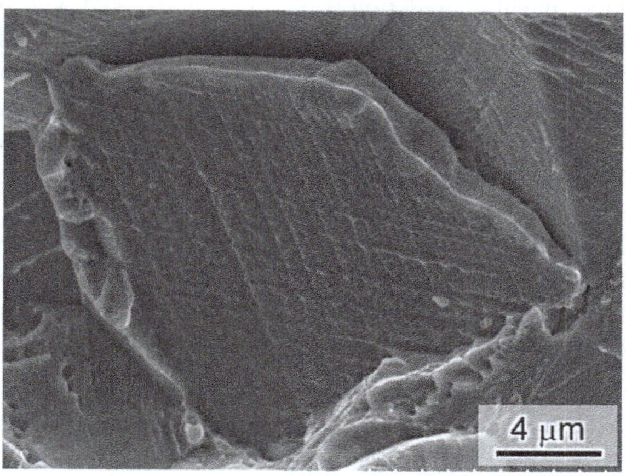

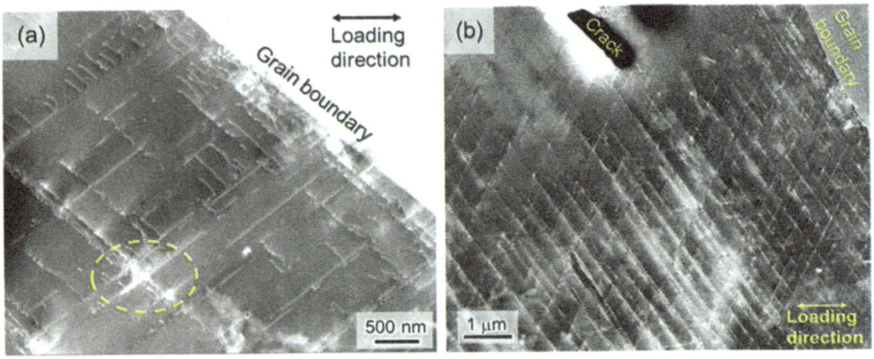

Fig. 1.12 **a** Contrast change at the intersection of planar dislocation arrays and **b** transgranular crack initiation, which corresponds to Figs. 1.9, 1.10 and 1.11 [21]. Reproduced with permission from *International Journal of Hydrogen Energy*, **45**, 10209–10218 (2020). Copyright 2020, Elsevier

1.4 Multiple Atomic Interactions: High-Entropy Alloy

The SRO effects on strengthening and dislocation patterns have also been reported in high-entropy alloys (HEAs), which have been noted as new types of high-strength and ductile materials [59–62]. A representative chemical composition of HEA is equiatomic CoCrFeMnNi with an FCC structure. The mechanical properties of FCC HEAs are similar to those of stable austenitic steels, yet showing relatively exceptional deformation behavior, perhaps owing to the presence of Co–Cr–Ni SRO. The SRO plays two important roles: tuning local stacking fault energy [63] and enhancing dislocation planarity [64]. The latter has a similar effect to that of Cr–N SRO. Because the quinary equiatomic HEA intrinsically shows low stacking fault energy, the additional effects of SRO result in distinct planar dislocation slips. Similar to other materials with high dislocation planarity, HEAs also show hydrogen-induced mechanical degradation (Fig. 1.13a) associated with plastic deformation near grain boundaries (Fig. 1.13b), causing the intergranular fracture (Fig. 1.13c when hydrogen was introduced by 100 MPa hydrogen gas charging [65, 66].[2]

In case a local stress near grain boundaries plays a critical role in intergranular fracture, grain refinement enables dramatic improvements in the ductility of metals, including austenitic steels, even when hydrogen was significantly introduced [69, 70]. Accordingly, the resistance to hydrogen-induced intergranular fracture can be improved by grain refinement in terms of tensile strength. As seen in Fig. 1.14a, the hydrogen-charged HEA with a grain size of 1.9 μm (FG800) shows comparable tensile strength to that without hydrogen charging, although the degradation of the ductility occurs even after grain refinement (Fig. 1.14b, c). The formation of precipitates such as the σ phase should be noted; however, it would degrade the strength

[2]When the hydrogen gas pressure was 15 MPa, the HEA showed no degradation of elongation and strength [67] .

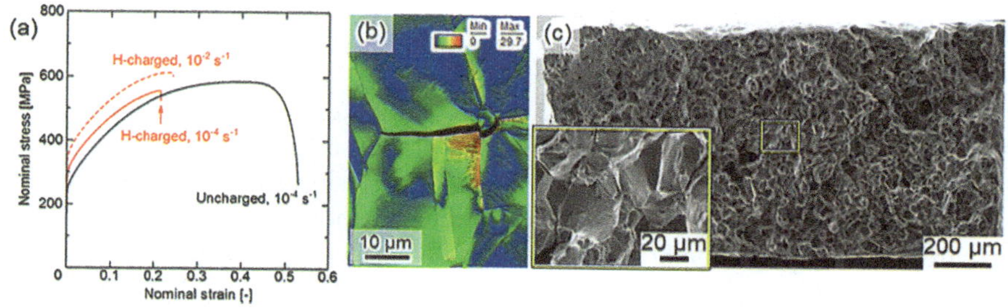

Fig. 1.13 **a** Engineering stress–strain curves of the equiatomic CoCrFeMnNi HEA. **b** Grain reference orientation deviation (GROD) map containing a crack and **(c)** fracture surface of the hydrogen-charged fractured specimen [66, 68]. Hydrogen was introduced at a hydrogen gas pressure of 100 MPa. GROD values indicate distribution of local plastic deformation. The inset indicates high-magnification of the region highlighted by the yellow square. Reproduced with permission from *Scripta Materialia*, **150**, 74–77 (2018). Copyright 2018, Elsevier

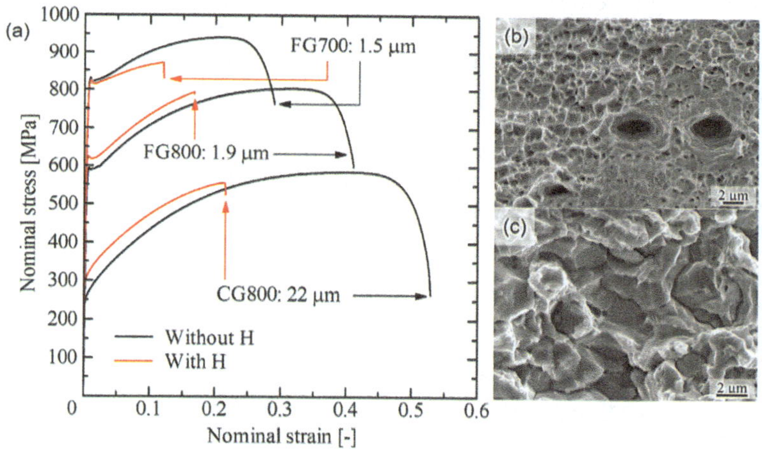

Fig. 1.14 **a** Engineering stress–strain curves of the equiatomic CoCrFeMnNi HEA with different grain sizes. **b**, **c** Fracture surfaces without and with hydrogen pre-charging [71]. Hydrogen was introduced at a hydrogen gas pressure of 100 MPa. Reproduced with permission from *International Journal of Hydrogen Energy*, **44**, 17163-17167 (2019). Copyright 2019, Elsevier

via hydrogen-assisted cracking at the interface between the FCC and σ phases. For instance, the degradation of work-hardening capability and tensile strength in the HEA with a grain size of 1.5 μm (FG700) results from the σ/γ interface cracking [71]. Therefore, decreasing the annealing temperature to obtain further grain-refined HEAs contains a risk of causing hydrogen embrittlement associated with the formation of the σ phase.

1.5 Summary

In this paper, we reviewed the potential effects of SRO on hydrogen embrittlement. Mn–C SRO induced dynamic strain aging and the subsequent enhancement of dislocation accumulation. Furthermore, the enhanced dislocation accumulation was localized in the form of PLC bands. Localized dislocation accumulation assists the occurrence of hydrogen embrittlement in high-Mn steels. Furthermore, as another type of SRO effect, Cr–N interactions were introduced. Cr–N SRO enhanced dislocation planarity, which subsequently caused microscopic stress concentrations as obstacles against planar dislocation motion. Stress concentration played a critical role by causing intergranular and transgranular cracking when hydrogen was introduced. Similar to the effect of Cr–N, Co–Cr–Ni SRO also enhanced dislocation planarity, subsequently assisting the occurrence of hydrogen-induced intergranular fractures. Hence, SRO possessed multiple disadvantageous effects on hydrogen embrittlement, although the SRO effect can realize an unconventional ductility-strength balance when hydrogen was not introduced. Since the majority of advanced FCC high-strength steels or alloys contain multiple components, we must carefully consider the SRO effects in avoiding unpredicted hydrogen-induced failures in terms of localization of plasticity.

Acknowledgements This work was financially supported by JSPS KAKENHI (JP16H06365 and JP20H02457) and the Japan Science and Technology Agency (JST) (Grant no. 20100113) under the Industry-Academia Collaborative R&D Program.

References

1. Zhang, L., Wen, M., Imade, M., Fukuyama, S., Yokogawa, K.: Effect of nickel equivalent on hydrogen gas embrittlement of austenitic stainless steels based on type 316 at low temperatures. Acta Mater. **56**(14), 3414–3421 (2008). https://doi.org/10.1016/j.actamat.2008.03.022
2. Omura, T., Nakamura, J., Hirata, H., Jotoku, K., Ueyama, M., Osuki, T., Terunuma, M.: Effect of surface hydrogen concentration on hydrogen embrittlement properties of stainless steels and Ni based alloys. ISIJ Int. **56**(3), 405–412 (2016). https://doi.org/10.2355/isijinternational.ISIJINT-2015-268
3. Izawa, C., Wagner, S., Deutges, M., Martin, M., Weber, S., Pargeter, R., Michler, T., Uchida, H.-H., Gemma, R., Pundt, A.: Relationship between hydrogen embrittlement and M_{d30} temperature: prediction of low-nickel austenitic stainless steel's resistance. Int. J. Hydr. Energy **44**(45), 25064–25075 (2019). https://doi.org/10.1016/j.ijhydene.2019.07.179
4. Koyama, M., Ogawa, T., Yan, D., Matsumoto, Y., Tasan, C.C., Takai, K., Tsuzaki, K.: Hydrogen desorption and cracking associated with martensitic trans-formation in Fe–Cr–Ni-Based austenitic steels with different carbon contents. Int. J. Hydr. Energy **42**(42), 26423–26435 (2017). https://doi.org/10.1016/j.ijhydene.2017.08.209
5. Teus, S.M., Shyvanyuk, V.N., Gavriljuk, V.G.: Hydrogen-induced $\gamma \rightarrow \varepsilon$ trans-formation and the role of ε-martensite in hydrogen embrittlement of austenit-ic steels. Mater. Sci. Eng. A **497**(1), 290–294 (2008). https://doi.org/10.1016/j.msea.2008.07.003
6. Chun, Y.S., Kim, J.S., Park, K.-T., Lee, Y.-K., Lee, C.S.: Role of ε martensite in tensile properties and hydrogen degradation of high-Mn steels. Mater. Sci. Eng. A **533**, 87–95 (2012). https://doi.org/10.1016/j.msea.2011.11.039

7. Koyama, M., Tasan, C.C., Tsuzaki, K.: Overview of metastability and compositional complexity effects for hydrogen-resistant iron alloys: inverse austenite stability effects. Eng. Fract. Mech. **214**, 123–133 (2019). https://doi.org/10.1016/j.engfracmech.2019.03.049
8. Koyama, M., Okazaki, S., Sawaguchi, T., Tsuzaki, K.: Hydrogen embrittlement susceptibility of Fe–Mn binary alloys with high Mn content: effects of stable and metastable ε-martensite, and Mn concentration. Metall. Mater. Trans. A **47**(6), 2656–2673 (2016). https://doi.org/10.1007/s11661-016-3431-9
9. Gibbs, P.J., Hough, P.D., Thürmer, K., Somerday, B.P., San Marchi, C., Zimmerman, J.A.: Stacking fault energy based alloy screening for hydrogen compatibility. JOM **72**(5), 1982–1992 (2020). https://doi.org/10.1007/s11837-020-04106-7
10. Symons, D.M.: Hydrogen embrittlement of Ni–Cr–Fe alloys. Metall. Mater. Trans. A **28**(3), 655–663 (1997). https://doi.org/10.1007/s11661-997-0051-4
11. Dastur, Y.N., Leslie, W.C.: Mechanism of work hardening in Hadfield manganese steel. Metall. Trans. A **12**(5), 749–759 (1981). https://doi.org/10.1007/BF02648339
12. Koyama, M., Shimomura, Y., Chiba, A., Akiyama, E., Tsuzaki, K.: Room-temperature blue brittleness of Fe-Mn-C austenitic steels. Scr. Mater. **141**, 20–23 (2017). https://doi.org/10.1016/j.scriptamat.2017.07.017
13. Owen, W.S., Grujicic, M.: Strain aging of austenitic Hadfield manganese steel. Acta Mater. **47**(1), 111–126 (1998). https://doi.org/10.1016/S1359-6454(98)00347-4
14. Lee, S.-J., Kim, J., Kane, S.N., Cooman, B.C.D.: On the origin of dynamic strain aging in twinning-induced plasticity steels. Acta Mater. **59**(17), 6809–6819 (2011). https://doi.org/10.1016/j.actamat.2011.07.040
15. Koyama, M., Akiyama, E., Tsuzaki, K.: Hydrogen-induced delayed fracture of a Fe-22Mn-0.6C steel pre-strained at different strain rates. Scr. Mater. **66**(11), 947–950 (2012). https://doi.org/10.1016/j.scriptamat.2012.02.040
16. Koyama, M., Akiyama, E., Tsuzaki, K.: Effects of static and dynamic strain aging on hydrogen embrittlement in TWIP steels containing Al. ISIJ Int. **53**(7), 1268–1274 (2013). https://doi.org/10.2355/isijinternational.53.1268
17. Michler, T., San Marchi, C., Naumann, J., Weber, S., Martin, M.: Hydrogen environment embrittlement of stable austenitic steels. Int. J. Hydr. Energy **37**(21), 16231–16246 (2012). https://doi.org/10.1016/j.ijhydene.2012.08.071
18. Koyama, M., Akiyama, E., Lee, Y.-K., Raabe, D., Tsuzaki, K.: Overview of hydrogen embrittlement in high-Mn steels. Int. J. Hydr. Energy **42**(17), 12706–12723 (2017). https://doi.org/10.1016/j.ijhydene.2017.02.214
19. Tomota, Y., Xia, Y., Inoue, K.: Mechanism of low temperature brittle fracture in high nitrogen bearing austenitic steels. Acta Mater. **46**(5), 1577–1587 (1998). https://doi.org/10.1016/S1359-6454(97)00350-9
20. Habib, K., Koyama, M., Tsuchiyama, T., Noguchi, H.: Visualization of dislocations through electron channeling contrast imaging at fatigue crack tip, interacting with pre-existing dislocations. Mater. Res. Lett. **6**(1), 61–66 (2018). https://doi.org/10.1080/21663831.2017.1392370
21. Koyama, M., Habib, K., Masumura, T., Tsuchiyama, T., Noguchi, H.: Gaseous hydrogen embrittlement of a Ni-free austenitic stainless steel containing 1 mass% nitrogen: effects of nitrogen-enhanced dislocation planarity. Int. J. Hydr. Energy **45**(16), 10209–10218 (2020). https://doi.org/10.1016/j.ijhydene.2020.02.014
22. Michler, T., Naumann, J.: Hydrogen embrittlement of Cr-Mn-N-austenitic stainless steels. Int. J. Hydr. Energy **35**(3), 1485–1492 (2010). https://doi.org/10.1016/j.ijhydene.2009.10.050
23. Karaman, I., Sehitoglu, H., Maier, H.J., Chumlyakov, Y.I.: Competing mechanisms and modeling of deformation in austenitic stainless steel single crystals with and without nitrogen. Acta Mater. **49**(19), 3919–3933 (2001). https://doi.org/10.1016/S1359-6454(01)00296-8
24. Oda, K., Kondo, N., Shibata, K.: X-ray absorption fine structure analysis of interstitial (C, N)-substitutional (Cr) complexes in austenitic stainless steels. ISIJ Int. **30**(8), 625–631 (1990). https://doi.org/10.2355/isijinternational.30.625
25. Koyama, M., Sawaguchi, T., Tsuzaki, K.: Overview of dynamic strain aging and associated phenomena in fe-mn-c austenitic steels. ISIJ Int. **58**(8), 1383–1395 (2018). https://doi.org/10.2355/isijinternational.ISIJINT-2018-237

26. Koyama, M., Sawaguchi, T., Lee, T., Lee, C.S., Tsuzaki, K.: Work hardening associated with ε-martensitic transformation, deformation twinning and dynamic strain aging in Fe-17Mn-0.6C and Fe-17Mn-0.8C TWIP steels. Mater. Sci. Eng. A **528**(24), 7310–7316 (2011). https://doi.org/10.1016/j.msea.2011.06.011
27. Oh, S.-K., Kilic, M.E., Seol, J.-B., Hong, J.-S., Soon, A., Lee, Y.-K.: The mechanism of dynamic strain aging for type A serrations in tensile flow curves of Fe-18Mn-0.55C (wt.%) twinning-induced plasticity steel. Acta Mater. **188**, 366–375 (2020). https://doi.org/10.1016/j.actamat.2020.02.020
28. Canadinc, D., Efstathiou, C., Sehitoglu, H.: On the negative strain rate sensitivity of Hadfield steel. Scr. Mater. **59**(10), 1103–1106 (2008). https://doi.org/10.1016/j.scriptamat.2008.07.027
29. Chen, L., Kim, H.-S., Kim, S.-K., De Cooman, B.C.: Localized Deformation due to Portevin–LeChatelier Effect in 18Mn& ndash;0.6C TWIP austenitic steel. ISIJ Int. **47**(12), 1804–1812 (2007). https://doi.org/10.2355/isijinternational.47.1804
30. Renard, K., Ryelandt, S., Jacques, P.J.: Characterisation of the Portevin-Le Châtelier effect affecting an austenitic TWIP steel based on digital image cor-relation. Mater. Sci. Eng. A **527**(12), 2969–2977 (2010). https://doi.org/10.1016/j.msea.2010.01.037
31. Yu, H.-Y., Lee, S.-M., Nam, J.-H., Lee, S.-J., Fabrègue, D., Park, M.-h., Tsuji, N., Lee, Y.-K.: Post-uniform elongation and tensile fracture mechanisms of Fe-18Mn-0.6C-xAl twinning-induced plasticity steels. Acta Mater. **131**, 435–444 (2017). https://doi.org/10.1016/j.actamat.2017.04.011
32. Najam, H., Koyama, M., Bal, B., Akiyama, E., Tsuzaki, K.: Strain rate and hydrogen effects on crack growth from a notch in a Fe-high-Mn steel containing 1.1 wt% solute carbon. Int. J. Hydr. Energy **45**(1), 1125–1139 (2020). https://doi.org/10.1016/j.ijhydene.2019.10.227
33. Hutchinson, B., Ridley, N.: On dislocation accumulation and work hardening in Hadfield steel. Scr. Mater. **55**(4), 299–302 (2006). https://doi.org/10.1016/j.scriptamat.2006.05.002
34. Hordon, M.J., Averbach, B.L.: X-ray measurements of dislocation density in deformed Copper and Aluminum single crystals. Acta Metall. **9**(3), 237–246 (1961). https://doi.org/10.1016/0001-6160(61)90073-6
35. Glarebrough, L.M., Hargreaves, M.E., West, G.W.: The density of dislocations in compressed copper. Acta Metall. **5**(12), 738–740 (1957). https://doi.org/10.1016/0001-6160(57)90076-7
36. Clarebrough, L.M., Hargreaves, M.E., West, G.W., Head, A.K., Mott, N.F.: The energy stored in fatigued metals. Proc. R. Soc. Lond. A Math Phys. Sci. **242**(1229), 160–166 (1957). https://doi.org/10.1098/rspa.1957.0164
37. Shintani, T., Murata, Y.: Evaluation of the dislocation density and dislocation character in cold rolled Type 304 steel determined by profile analysis of X-ray diffraction. Acta Mater. **59**(11), 4314–4322 (2011). https://doi.org/10.1016/j.actamat.2011.03.055
38. Dini, G., Ueji, R., Najafizadeh, A., Monir-Vaghefi, S.M.: Flow stress analysis of TWIP steel via the XRD measurement of dislocation density. Mater. Sci. Eng. A **527**(10), 2759–2763 (2010). https://doi.org/10.1016/j.msea.2010.01.033
39. Tuğuca, I.B., Koyama, M., Bal, B., Canadinc, D., Akiyama, E., Tsuzaki, K.: High-concentration carbon assists plasticity-driven hydrogen embrittlement in a Fe-high Mn steel with a relatively high stacking fault energy. Mater. Sci. Eng. A **717**, 78–84 (2018). https://doi.org/10.1016/j.msea.2018.01.087
40. Tuğluca, I.B., Koyama, M., Shimomura, Y., Bal, B., Canadinc, D., Akiyama, E., Tsuzaki, K.: Lowering strain rate simultaneously enhances carbon- and hydrogen-induced mechanical degradation in an Fe-33Mn-1.1C Steel. Metall. Mater. Trans. A **50**(3), 1137–1141 (2019). https://doi.org/10.1007/s11661-018-5080-7
41. Bouaziz, O., Allain, S., Scott, C.P., Cugy, P., Barbier, D.: High manganese austenitic twinning induced plasticity steels: a review of the microstructure properties relationships. Curr. Opin. Solid State Mater. Sci. **15**(4), 141–168 (2011). https://doi.org/10.1016/j.cossms.2011.04.002
42. Chowdhury, P., Canadinc, D., Sehitoglu, H.: On deformation behavior of Fe-Mn based structural alloys. Mater. Sci. Eng. R **122**, 1–28 (2017). https://doi.org/10.1016/j.mser.2017.09.002
43. Liang, Z.Y., Wang, X., Huang, W., Huang, M.X.: Strain rate sensitivity and evolution of dislocations and twins in a twinning-induced plasticity steel. Acta Mater. **88**, 170–179 (2015). https://doi.org/10.1016/j.actamat.2015.01.013

44. Koyama, M., Akiyama, E., Sawaguchi, T., Raabe, D., Tsuzaki, K.: Hydrogen-induced cracking at grain and twin boundaries in an Fe-Mn-C austenitic steel. Scr. Mater. **66**(7), 459–462 (2012). https://doi.org/10.1016/j.scriptamat.2011.12.015
45. Koyama, M., Akiyama, E., Tsuzaki, K., Raabe, D.: Hydrogen-assisted failure in a twinning-induced plasticity steel studied under in situ hydrogen charging by electron channeling contrast imaging. Acta Mater. **61**(12), 4607–4618 (2013). https://doi.org/10.1016/j.actamat.2013.04.030
46. Ryu, J.H., Kim, S.K., Lee, C.S., Suh, D.-W., Bhadeshia, H.K.D.H.: Effect of aluminium on hydrogen-induced fracture behaviour in austenitic Fe–Mn–C steel. Proc. R. Soc. A: Mathematical, Physical and Engineering Sciences **469**(2149), 20120458 (2013). https://doi.org/10.1098/rspa.2012.0458
47. Koyama, M., Bashir, A., Rohwerder, M., Merzlikin, S.V., Akiyama, E., Tsuzaki, K., Raabe, D.: Spatially and kinetically resolved mapping of hydrogen in a twinning-induced plasticity steel by use of scanning kelvin probe force microscopy. J. Electrochem. Soc. **162**(12), C638–C647 (2015). https://doi.org/10.1149/2.0131512jes
48. So, K.H., Kim, J.S., Chun, Y.S., Park, K.-T., Lee, Y.-K., Lee, C.S.: Hydrogen delayed fracture properties and internal hydrogen behavior of a Fe-18Mn-1.5Al-0.6C TWIP Steel. ISIJ Int. **49**(12), 1952–1959 (2009). https://doi.org/10.2355/isijinternational.49.1952
49. Du, Y.A., Ismer, L., Rogal, J., Hickel, T., Neugebauer, J., Drautz, R.: First-principles study on the interaction of H interstitials with grain boundaries in α- and γ-Fe. Phys. Rev. B **84**(14), 144121 (2011). https://doi.org/10.1103/PhysRevB.84.144121
50. Mahajan, S., Chin, G.Y.: Twin-slip, twin-twin and slip-twin interactions in Co-8 wt.% Fe alloy single crystals. Acta Metall. **21**(2), 173–179 (1973). https://doi.org/10.1016/0001-6160(73)90059-X
51. Rémy, L.: Twin-slip interaction in f.c.c. crystals. Acta Metall. **25**(6), 711–714 (1977). https://doi.org/10.1016/0001-6160(77)90013-X
52. Wang, Y.B., Sui, M.L.: Atomic-scale in situ observation of lattice dislocations passing through twin boundaries. Appl. Phys. Lett. **94**(2), 021909 (2009). https://doi.org/10.1063/1.3072801
53. Bal, B., Koyama, M., Gerstein, G., Maier, H.J., Tsuzaki, K.: Effect of strain rate on hydrogen embrittlement susceptibility of twinning-induced plasticity steel pre-charged with high-pressure hydrogen gas. Int. J. Hydr. Energy **41**(34), 15362–15372 (2016). https://doi.org/10.1016/j.ijhydene.2016.06.259
54. Koyama, M., Rohwerder, M., Tasan, C.C., Bashir, A., Akiyama, E., Takai, K., Raabe, D., Tsuzaki, K.: Recent progress in microstructural hydrogen mapping in steels: quantification, kinetic analysis, and multi-scale characterisation. Mater. Sci. Technol. **33**(13), 1481–1496 (2017). https://doi.org/10.1080/02670836.2017.1299276
55. Masumura, T., Seto, Y., Tsuchiyama, T., Kimura, K.: Work-hardening mechanism in high-nitrogen austenitic stainless steel. Mater. Trans. **61**(4), 678–684 (2020). https://doi.org/10.2320/matertrans.H-M2020804
56. Terazawa, Y., Ando, T., Tsuchiyama, T., Takaki, S.: Relationship between work hardening behaviour and deformation structure in Ni-free high nitro-gen austenitic stainless steels. Steel Res. Int. **80**(7), 473–476 (2009). https://doi.org/10.2374/sri09sp039
57. Tsuchiyama, T., Fujii, Y., Terazawa, Y., Nakashima, K., Ando, T., Takaki, S.: Factors inducing intergranular fracture in nickel-free high nitrogen austen-itic stainless steel produced by solution nitriding. ISIJ Int. **48**(6), 861–867 (2008). https://doi.org/10.2355/isijinternational.48.861
58. Kubota, S., Xia, Y., Tomota, Y.: Work-hardening behavior and evolution of dislocation-microstructures in high-nitrogen bearing austenitic steels. ISIJ Int. **38**(5), 474–481 (1998). https://doi.org/10.2355/isijinternational.38.474
59. Yeh, J.-W., Chen, S.-K., Lin, S.-J., Gan, J.-Y., Chin, T.-S., Shun, T.-T., Tsau, C.-H., Chang, S.-Y.: Nanostructured high-entropy alloys with multiple principal elements: novel alloy design concepts and outcomes. Adv. Eng. Mater. **6**(5), 299–303 (2004). https://doi.org/10.1002/adem.200300567
60. Cantor, B.: Multicomponent and high entropy alloys. Entropy **16**(9), 4749 (2014)

61. Gludovatz, B., Hohenwarter, A., Catoor, D., Chang, E.H., George, E.P., Ritchie, R.O.: A fracture-resistant high-entropy alloy for cryogenic applications. Science **345**(6201), 1153–1158 (2014). https://doi.org/10.1126/science.1254581
62. Yao, M.J., Pradeep, K.G., Tasan, C.C., Raabe, D.: A novel, single phase, non-equiatomic FeMnNiCoCr high-entropy alloy with exceptional phase stability and tensile ductility. Scr. Mater. **72–73**, 5–8 (2014). https://doi.org/10.1016/j.scriptamat.2013.09.030
63. Ding, J., Yu, Q., Asta, M., Ritchie, R.O.: Tunable stacking fault energies by tailoring local chemical order in CrCoNi medium-entropy alloys. Proc. Natl. Acad. Sci. USA **115**(36), 8919–8924 (2018). https://doi.org/10.1073/pnas.1808660115
64. Zhang, R., Zhao, S., Ding, J., Chong, Y., Jia, T., Ophus, C., Asta, M., Ritchie, R.O., Minor, A.M.: Short-range order and its impact on the CrCoNi medium-entropy alloy. Nature **581**(7808), 283–287 (2020). https://doi.org/10.1038/s41586-020-2275-z
65. Nygren, K.E., Bertsch, K.M., Wang, S., Bei, H., Nagao, A., Robertson, I.M.: Hydrogen embrittlement in compositionally complex FeNiCoCrMn FCC solid solution alloy. Curr. Opin. Solid State Mater. Sci. **22**(1), 1–7 (2018). https://doi.org/10.1016/j.cossms.2017.11.002
66. Ichii, K., Koyama, M., Tasan, C.C., Tsuzaki, K.: Comparative study of hydro-gen embrittlement in stable and metastable high-entropy alloys. Scr. Mater. **150**, 74–77 (2018). https://doi.org/10.1016/j.scriptamat.2018.03.003
67. Zhao, Y., Lee, D.-H., Seok, M.-Y., Lee, J.-A., Phaniraj, M.P., Suh, J.-Y., Ha, H.-Y., Kim, J.-Y., Ramamurty, U., Jang, J.-i.: Resistance of CoCrFeMnNi high-entropy alloy to gaseous hydrogen embrittlement. Scr. Mater. **135**, 54–58 (2017). https://doi.org/10.1016/j.scriptamat.2017.03.029
68. Ichii, K., Koyama, M., Tasan, C.C., Tsuzaki, K.: Localized plasticity and associated cracking in stable and metastable high-entropy alloys pre-charged with hydrogen. Procedia Struct. Integr. **13**, 716–721 (2018). https://doi.org/10.1016/j.prostr.2018.12.119
69. Park, I.-J., Lee, S.-M., Jeon, H.-h., Lee, Y.-K.: The advantage of grain refinement in the hydrogen embrittlement of Fe-18Mn-0.6C twinning-induced plasticity steel. Corros. Sci. **93**, 63–69 (2015). https://doi.org/10.1016/j.corsci.2015.01.012
70. Zan, N., Ding, H., Guo, X., Tang, Z., Bleck, W.: Effects of grain size on hydrogen embrittlement in a Fe-22Mn-0.6C TWIP steel. Int. J. Hydr. Energy **40**(33), 10687–10696 (2015). https://doi.org/10.1016/j.ijhydene.2015.06.112
71. Koyama, M., Ichii, K., Tsuzaki, K.: Grain refinement effect on hydrogen embrittlement resistance of an equiatomic CoCrFeMnNi high-entropy alloy. Int. J. Hydr. Energy **44**(31), 17163–17167 (2019). https://doi.org/10.1016/j.ijhydene.2019.04.280

Chapter 2
Acoustic Anisotropy and Hydrogen Concentrations During Thermomechanical Loading of Single-Crystal Nickel-Based Superalloys

Alexander K. Belyaev, Aliya R. Galyautdinova, Vladimir A. Polyanskiy, Artem S. Semenov, Dmitry A. Tretyakov, and Yuriy A. Yakovlev

Abstract Investigations of hydrogen concentrations and acoustic anisotropy in single-crystal nickel-based specimens were carried out. The growth of creep and thermal fatigue cracks for different modes of thermomechanical loading was studied. Experimental data and theoretical estimates of acoustic anisotropy in specimens with crystallographic direction $\langle 011 \rangle$ of face-centered cubic lattice were obtained. It was found that anisotropy of elastic modulus provides a main contribution to acoustic anisotropy in the case of single-crystal alloys. Measurements of hydrogen concentrations revealed its accumulation after thermomechanical loading in a weakly bound state along edges of specimens to the level of 4 ppm. It indicates a significant degradation of mechanical properties and the presence of developed hydrogen embrittlement. The obtained results allow one to develop an integrated approach for estimating the residual life of single-crystal structures by analyzing hydrogen concentrations and acoustic anisotropy parameters.

Keywords Single Crystal Superalloy · Creep · Thermal fatigue · Acoustoelastic effect · Hydrogen diffusion · Elastic sound waves

---

A. K. Belyaev · A. R. Galyautdinova · V. A. Polyanskiy · A. S. Semenov · D. A. Tretyakov (✉) · Y. A. Yakovlev
Peter the Great St. Petersburg Polytechnic University (SPbPU), St. Petersburg, Russia
e-mail: tretyakov_da@spbstu.ru

A. K. Belyaev
e-mail: vice.ipme@gmail.com

A. R. Galyautdinova
e-mail: aliyagalyautdinova@gmail.com

V. A. Polyanskiy
e-mail: vapol@mail.ru

A. S. Semenov
e-mail: semenov.artem@googlemail.com

Y. A. Yakovlev
e-mail: yura.yakovlev@gmail.com

V. A. Polyanskiy and A. K. Belyaev (eds.), *Advances in Hydrogen Embrittlement Study*, Advanced Structured Materials 143, https://doi.org/10.1007/978-3-030-66948-5_2

2.1 Inroduction

Estimation of residual life of power equipment is one of the urgent tasks of technical diagnostics. Blades are critical elements of modern gas turbine engines. They are grown from heat-resistant materials according to the Bridgman method [1]. It is known that the destruction of polycrystalline blades occurs along grain boundaries under influence of centrifugal forces. The exclusion of grain boundaries by using single-crystal materials with directional crystallization allows one to increase strength and durability of gas turbine blades.

There are strict requirements for high-temperature strength, thermomechanical fatigue, and high-temperature creep of gas turbine blades at operating temperatures above 1000 °C. At the same time, there are no substantiated experimental criteria for thermal fatigue strength of single-crystal materials under complex thermomechanical loading. Also, there are no universal approaches for detecting defects caused by hydrogen embrittlement, thermal fatigue, and creep. For this reason, development of modern methods of non-destructive testing is necessary to ensure safe operation of power plants.

The use of acoustic anisotropy parameter for estimating current state of industrial structures has shown its effectiveness [2–5]. The observed stable correlations between acoustic anisotropy of shear waves, concentrations of dissolved hydrogen, and damage measures [6] suggest new possibilities for detecting hydrogen embrittlement in metals. In particular, the skin effect of hydrogen accumulation [7] and the surface effect of acoustic anisotropy [8] were discovered for polycrystalline materials. The fundamental problem of mutual influence of hydrogen concentrations, microdefects, and changes on acoustic parameters was posed earlier [6].

The aim of this work is to study distributions of acoustic anisotropy parameter and dissolved hydrogen concentrations in single-crystal heat-resistant specimens after thermal fatigue tests.

2.2 Acoustoelastic Effect in Single-Crystal Materials

Acoustoelasticity as a property of solids has been known for a long time [9]. Bridgman studied the influence of high compressive pressures on elastic modulus of metals [10, 11]. Birch [12] and Biot [13] solved the problem of elastic wave propagation in initially stressed isotropic media. Hughes and Kelly [14] obtained equations describing the propagation of longitudinal and shear waves in a nonlinear elastic material according to Murnaghan's theory of finite deformations [15]. They laid foundations of the acoustoelasticity theory for the case of uniaxial stress state [15]. The theory of acoustoelasticity describes the influence of mechanical stresses on velocities of elastic waves propagation in solids. Benson and Raelson called it the acoustoelastic effect [16] by analogy with the photoelastic effect.

The general theory of acoustoelasticity was developed by Toupin and Bernstein [17]. They obtained equations for velocities of elastic waves in a nonlinear elastic material with arbitrary symmetry. Fukuoka [18], based on results obtained by Tokuoka [19] and Okada [20], found the acoustoelasticity equation (2.1) for an isotropic material:

$$\frac{V_{T_1} - V_{T_2}}{V_{T_0}} = \alpha + C_A(\sigma_1 - \sigma_2), \tag{2.1}$$

where V_{T_0} is shear wave velocity in unstressed material, V_{T_1}, V_{T_2} are shear wave velocities polarized along axes of principal stress σ_1 and σ_2, α is the coefficient of texture-induced anisotropy, C_A is the acoustoelasticity constant of material. The value of acoustic birefringence $\Delta a = \Delta V / V_0$ (2.1) of elastic shear waves is called the acoustic anisotropy. The coefficient α and the constant C_A can be obtained from Eqs. (2.2) and (2.3):

$$C_A = \frac{1}{2\mu}(1 + \frac{\nu_3}{\mu}), \tag{2.2}$$

$$\alpha = \frac{C_{55} - C_{44}}{2\mu}, \tag{2.3}$$

where μ is the second-order elastic constant, ν_3 is the third-order elastic constant in the notation of Toupin and Bernstein [17], C_{44} and C_{55} are the elastic constants in the notation of Voigt.

Practical results of acoustoelasticity investigations are mainly related with determination of the third-order elastic constants for different materials [21–24] and estimation of biaxial stresses in weakly anisotropic structures [25–30]. Technology for measuring biaxial stresses in industrial structures made from polycrystalline materials was developed in [31]. Equipment for acoustic anisotropy analysis in extended structures such as pipelines was developed in [32, 33].

At the same time, acoustic anisotropy is a more complex characteristic of the current state of structures [34, 35]. Recent studies indicate its relationship with surface damage [8], accumulated due to plastic deformation [36–39], corrosion [40], hydrogen embrittlement [6], and other reasons leading to destruction of material.

Scientific results related to acoustoelastic effect in single-crystal materials are limited to theoretical description of wave propagation in nonlinearly elastic media with different crystal symmetry. Seeger and Buck [41] determined the third-order elastic constants for single-crystal germanium. Hikata [42, 43] investigated the attenuation of ultrasonic waves using dislocation theory. Einspruch and Manning [44] calculated the third-order elastic constants for tetragonal, orthorhombic, and hexagonal crystal lattices. Generalized equations describing wave propagation in stressed crystals were obtained by Thurston [45].

Theoretical study of acoustoelastic effect in undamaged elastically loaded crystals was carried out by Tokuoka and Saito [19]. However, the relationships obtained in [19] cannot be applied to real single-crystal structures grown according to the industrial technology of directional crystallization.

Experiments on rotor blades of high-pressure turbocharger were carried out by Nikitina [46, 47]. Acoustic anisotropy was measured on heat-resistant blades before and after testing during the warranty period of 9000 h. The maximum value of acoustic anisotropy observed on single-crystal specimens made of heat-resistant alloy AISI 431 was equal to 1.65% [47].

The observed values of acoustic anisotropy [47] are explained as the result of thermal fatigue accumulation in blades during their exploitation. At the same time, acoustic anisotropy should be related to internal anisotropy of mechanical properties in accordance with formula (2.1). Anisotropy of elastic moduli is due to crystallographic direction of specimen growth [17].

It is necessary to study distributions of acoustic anisotropy and accumulated hydrogen concentrations in single-crystal heat-resistant specimens with different crystallographic orientation.

2.3 Experimental Studies of Acoustic Anisotropy and Hydrogen Concentrations in Single-Crystal Nickel-Based Alloys

Investigations of hydrogen concentrations and acoustic anisotropy parameter were carried out on plane single-crystal specimens made of two different nickel-based alloys, which were modifications of EPM102 and CMSX-10 heat-resistant alloys.

Nickel-based alloys have a dendritic-cellular structure with a face-centered cubic lattice (FCC) oriented along different crystallographic directions. They have improved properties of long-term strength, ductility, heat resistance, and gas corrosion resistance due to nickel alloying. It allows one to use nickel-based alloys in the production of gas turbine blades.

Single-crystal specimens (in Fig. 2.1) were tested in the case of creep cracks growth and thermal fatigue accumulation after different modes of thermomechanical loading [48, 49]. Figure 2.1 shows the halves of destroyed specimens after testing. Large specimens with a size of $35\times35\times7$ mm were made of Alloy 1. Their schematic view is shown in Fig. 2.2. Smaller specimens from Fig. 2.1 with a size of $31\times31\times6$ mm were made of Alloy 2. Their schematic view is similar to specimens shown in Fig. 2.2.

Investigations were carried out on specimens grown along different crystallographic directions. In addition to structural anisotropy, the influence of creep cracks and thermal fatigue was taken into account. For this reason, the conditions of thermomechanical loading at operating cycle temperatures of 20 °C and 1000 °C were considered.

Fig. 2.1 Single-crystal specimens after thermal fatigue tests with different temperature loading conditions

Fig. 2.2 Schematic view of single-crystal specimens from Alloy 1

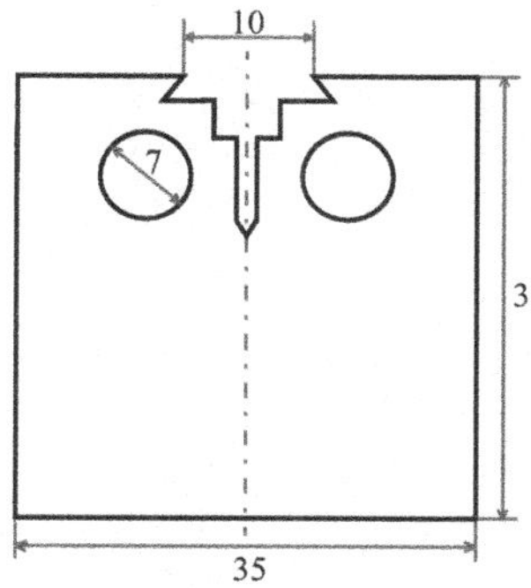

Fig. 2.3 Propagation of shear waves V_1, V_2 with mutually perpendicular polarization

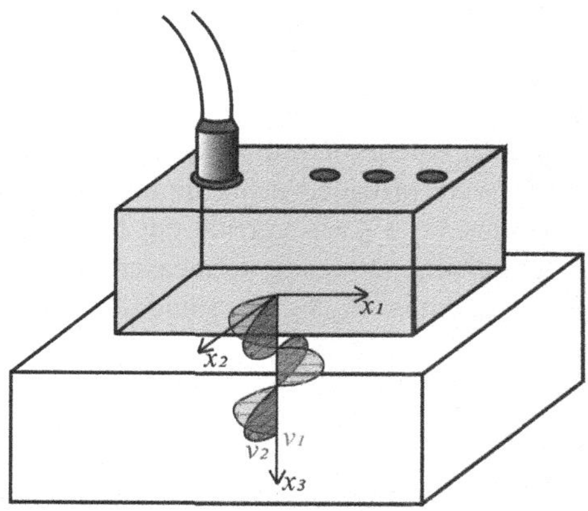

Ultrasonic measurements were carried out using acoustic sensor (in Fig. 2.3) of industrial acoustic anisotropy analyzer (in Fig. 2.4) [47]. Generation of shear wave pulses was carried out using piezoelectric transducers with a wave frequency of 5 MHz.

Fig. 2.4 Industrial acoustic anisotropy analyzer with ultrasonic sensor

Table 2.1 Experimental values of acoustic anisotropy for single-crystal heat-resistant specimens after thermal fatigue tests with different modes of thermomechanical loading

Nickel-based alloy and specimen number	Crystallographic direction	Load, kN	Temperature, °C	Acoustic anisotropy $\|\Delta a\|$,%
Alloy 1, Specimen No.1	⟨011⟩	5	20	47.79
Alloy 1, Specimen No.2	⟨011⟩	3	1000	46.81
Alloy 1, Specimen No.3	⟨001⟩	4	1000	4.23
Alloy 1, Specimen No.4	⟨001⟩	5	1000	0.46
Alloy 2, Specimen No.5	⟨011⟩	3	20	33.89
Alloy 2, Specimen No.6	⟨011⟩	3	900	33.65

Measurements of time delays t_1, t_2 between multiple reflected packets of ultrasonic pulses were provided using digital software of device. Calculation of shear wave velocities V_1, V_2 was carried out by direct measurements of average thickness in zones of time delays t_1, t_2 measurement. Acoustic anisotropy Δa was calculated using velocities V_1, V_2 according to Eq. (2.4):

$$\Delta a = \frac{v_1 - v_2}{(v_1 + v_2)/2} \tag{2.4}$$

The experimental values of acoustic anisotropy modulus $|\Delta a|$ for six single-crystal specimens from two different alloys are presented in Table 2.1. Table 2.1 shows the crystallographic directions of specimen growth, temperature conditions, and extreme loads in one loading cycle.

Investigation of hydrogen concentrations was carried out on three specimens No.7, No.8, and No.9 from Alloy 1 according to the vacuum heating method [50]. The schematic view of specimens is the same as for large specimens from Alloy 1 shown in Fig. 2.2. Specimen No.7 was cut from material in its initial state. Specimens

Fig. 2.5 Industrial hydrogen analyzer

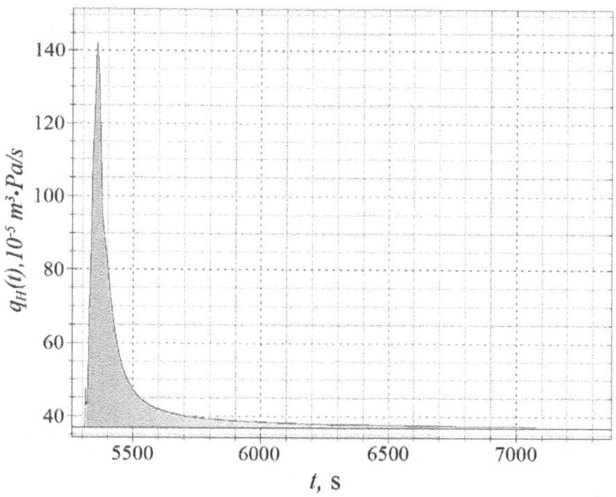

Fig. 2.6 Experimental curve $q_H(t)$ of hydrogen extraction for Alloy 1

No.8 and No.9 were subjected to thermal fatigue tests until destruction. Hydrogen concentrations were measured in two areas located near the center and at the edge of each specimen.

Measurements were carried out using an industrial hydrogen analyzer (in Fig. 2.5). The optimum temperatures of hydrogen extraction are equal to 530 °C and 800 °C for Alloy 1. The binding energies of hydrogen were estimated using the multichannel diffusion model [7]. Comparison of the experimental extraction curves (in Fig. 2.6) with results of numerical simulation revealed main hydrogen energy states, which were equal to 0.5, 1.2, and 1.5 eV.

The study revealed hydrogen accumulation at the edges of specimens No.8 and No.9 after thermal fatigue tests. The concentration of weakly bound hydrogen increased by more than 10 times and reached the critical level of 4 ppm.

2.4 Discussion

Experimental studies show that initial acoustic anisotropy in case of single-crystal specimens significantly exceed values of Δa observed for isotropic materials. The maximum acoustic anisotropy level of isotropic polycrystalline metal don't exceed 3% [5]. At the same time, observed values of acoustic anisotropy turn out to be close to 50% in the case of single-crystal specimens (see Table 2.1). It indicates a strong influence of anisotropy of elastic modulus on values of acoustic anisotropy parameter Δa.

The generalized Hooke's law for anisotropic material has the form (2.5):

$$\boldsymbol{\sigma} = {}^4\mathbf{C} \cdot\cdot\, \boldsymbol{\varepsilon}, \tag{2.5}$$

where elastic modulus tensor ${}^4\mathbf{C}$ for material with cubic symmetry is determined by relation (2.6):

$${}^4\mathbf{C} = \lambda \mathbf{1} \otimes \mathbf{1} + \mu(\mathbf{1}\overline{\otimes}\mathbf{1} + \mathbf{1}\underline{\otimes}\mathbf{1}) + \alpha \sum_{k=1}^{3} \mathbf{m}_k \otimes \mathbf{m}_k \otimes \mathbf{m}_k \otimes \mathbf{m}_k. \tag{2.6}$$

The symbols of direct and indirect dyadic multiplication $(\mathbf{A} \otimes \mathbf{B})_{ijkl} = A_{ij} B_{kl}$, $(\mathbf{A}\overline{\otimes}\mathbf{B})_{ijkl} = A_{ik} B_{jl}$, $(\mathbf{A}\underline{\otimes}\mathbf{B})_{ijkl} = A_{il} B_{jk}$ were used in (2.6). The unit vectors $\mathbf{m}_k$ correspond to mutually orthogonal directions $\mathbf{m}_1 = [100], \mathbf{m}_2 = [010], \mathbf{m}_3 = [001]$ of crystal lattice. The basic unit vectors $\mathbf{e}_1, \mathbf{e}_2, \mathbf{e}_3$ of the laboratory coordinate system do not coincide with the unit vectors $\mathbf{m}_1$, $\mathbf{m}_2$, $\mathbf{m}_3$ in general case. In relation (2.6), three independent material constants λ, μ, α were used. The first and last constants were calculated through Young's modulus and Poisson's ratio using Eqs. (2.7):

$$\lambda = \frac{\nu E}{(1+\nu)(1-2\nu)}, \qquad \alpha = \frac{E}{1+\nu} - 2\mu. \tag{2.7}$$

The positive definiteness of elastic modulus tensor ${}^4\mathbf{C}$ (2.6) is satisfied for $\mu > 0$ and $\lambda + 2\mu + \alpha > 0$. It corresponds to conditions for an isotropic material $E > 0$ and $-1 < \nu \leq 1/2$.The parameter α can take both positive or negative values.

Phase velocities v of plane monochromatic wave propagation in the direction $\mathbf{N}(|\mathbf{N}| = 1)$ (2.8)

$$\mathbf{u}(\mathbf{r}, t) = \mathbf{A} \exp ik(\mathbf{N} \cdot \mathbf{r} - vt) \tag{2.8}$$

characterized by elastic modulus tensor ${}^4\mathbf{C}$ and density ρ. In this case, the phase velocities are determined from the solution of Eq. (2.9):

$$\det(\mathbf{N}\cdot{}^4\mathbf{C}\cdot\mathbf{N}-\rho\nu^2\mathbf{1})=0, \tag{2.9}$$

where (2.9) is the condition for determining eigenvalues of acoustic tensor $\mathbf{Q}=\mathbf{N}\cdot{}^4\mathbf{C}\cdot\mathbf{N}$.

Acoustic tensor $\mathbf{Q}$ can be specified in the form (2.10) for a material with cubic symmetry:

$$\mathbf{Q}=\mathbf{N}\cdot{}^4\mathbf{C}\cdot\mathbf{N}=(\lambda+\mu)\mathbf{N}\otimes\mathbf{N}+\mu\mathbf{1}+\alpha\sum_{k=1}^{3}(\mathbf{N}\cdot\mathbf{m}_k)^2\,\mathbf{m}_k\otimes\mathbf{m}_k\,. \tag{2.10}$$

The acoustic tensor $\mathbf{Q}$ can be obtained from (2.10) by wave propagation in the direction $\mathbf{N}=\mathbf{e}_3$. It will have the following form for the laboratory coordinate system (2.11):

$$\mathbf{Q}=(\lambda+\mu)\mathbf{e}_3\otimes\mathbf{e}_3+\mu\mathbf{1}+\alpha\mathbf{e}_i\otimes\mathbf{e}_j\sum_{k=1}^{3}\alpha_{3k}^2\alpha_{ik}\alpha_{jk}\,, \tag{2.11}$$

where $\alpha_{ij}=\mathbf{e}_i\cdot\mathbf{m}_j$. The components of the fourth rank tensor of elastic modulus ${}^4\mathbf{C}$ change according to rule $C_{ijkl}=\alpha_{ip}\alpha_{jr}\alpha_{ks}\alpha_{lt}C_{prst}$.

Nonzero values of acoustic anisotropy appear in the case of specimen axes deviation from crystallographic axes. It agrees with Okada equations [20] obtained in linear theory of acoustoelasticity. The orientational dependence of acoustic anisotropy $\Delta a(\phi,\theta)$ obtained from the solution of Eq. (2.9) is shown in Fig. 2.7b.

The maximum values of acoustic anisotropy Δa (2.4) among all possible crystal orientations are observed to twelve directions $\langle 011\rangle$ (in Fig. 2.7b). They do not correspond to the maximum values of elasticity modulus realized for eight directions $\langle 111\rangle$ (in Fig. 2.7a). An abnormally high level of initial acoustic anisotropy reaches 46% for unstressed and undamaged specimens No.1 and No.2 from Alloy 1.

The roots of Eq. (2.9) for direction [101] have the following form (2.12):

$$\begin{cases}\nu_3=\sqrt{(\lambda+2\mu+\alpha/2)/\rho}\\ \nu_2=\sqrt{(\mu+\alpha/2)/\rho}\\ \nu_1=\sqrt{\mu/\rho}.\end{cases}, \tag{2.12}$$

The maximum values of acoustic anisotropy Δa for 12 directions $\langle 011\rangle$ are determined by the relation (2.13):

$$\Delta a=\frac{\alpha/\mu}{2+\alpha/2\mu+2\sqrt{1+\alpha/2\mu}}\,. \tag{2.13}$$

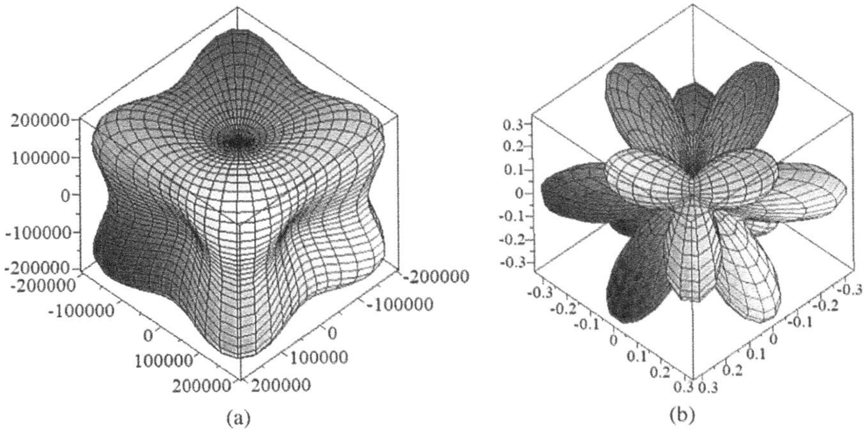

Fig. 2.7 Orientation dependence of Young's modulus E_z **a** and acoustic anisotropy Δa **b** for single crystals with cubic symmetry

In accordance with (2.13), acoustic anisotropy Δa is determined only by the value $\alpha/2\mu$ and does not depend on the Poisson's ratio ν and density ρ.

According to Eq. (2.13), acoustic anisotropy $|\Delta a|$ for Alloy 1 with the elastic modulus $E_{[101]} = 130000$ MPa, $\nu = 0.39$, $\mu = 119000$ MPa at the temperature of 20 °C is equal to $|\Delta a| = 45.9\%$. This result correlates with the experimentally observed value for specimen No.1 $|\Delta a| = 47.8\%$ (see Table 2.1). The acoustic anisotropy value $|\Delta a|$ for Alloy 2 with elastic modulus $E_{[101]} = 137000$ MPa, $\nu = 0.395$, $\mu = 125000$ MPa at the temperature of 20 °C is equal to $|\Delta a| = 45.8\%$. This value is higher than experimentally observed acoustic anisotropy for specimen No.5 $|\Delta a| = 33.9\%$. It is due to deviation of the specimen orientation from crystallographic direction [101]. The theoretical value of acoustic anisotropy decreases by 7% in the case of deviation of specimen axial orientation from crystallographic direction [101] by the angle 10°. The values of acoustic anisotropy for specimens No.3 and No.4 with direction [100] are lower than for direction [101] by an order of magnitude.

Residual deformations and damage also impact on acoustic anisotropy [4, 5, 8]. Their contribution to integral value of acoustic anisotropy is equal to 0.1-0.3% for steel and aluminum alloys [5]. The values of acoustic anisotropy related to residual deformations and damage have the same level in single crystals as for polycrystalline materials (see Table 2.1).

Investigations of hydrogen concentrations were carried out according to the vacuum heating method [7]. Hydrogen accumulation in a weakly bound diffusion state was observed at the edges of specimens. The accumulated hydrogen concentrations in specimens No.8 and No.9 were equal to 4 ppm. It is known that the maximum hydrogen concentration in nickel-based alloys at atmospheric pressure is equal to 5 ppm. Thus, specimens after thermal fatigue tests accumulated critical hydrogen concentrations. This diagnostic sign indicates a significant degradation of mechanical properties of single-crystal specimens. It leads to brittle fracture of material due to hydrogen embrittlement [7].

2.5 Conclusions

The study of acoustic anisotropy Δa and hydrogen concentrations of single-crystal heat-resistant specimens from nickel-based alloys was carried out. The values of Δa for specimens with crystallographic direction $\langle 011\rangle$ of a cubic crystal system exceeded acoustic anisotropy observed in polycrystalline materials and single-crystal specimens with direction $\langle 001\rangle$ by order of magnitude. It was found that anisotropy of elastic modulus can provide main contribution to the value of acoustic anisotropy in case of single-crystal materials. The maximum levels of Δa were observed for 12 directions $\langle 011\rangle$.

The results of ultrasonic measurements showed that accumulated damage and residual deformations in single-crystal specimens have a weak influence on acoustic anisotropy Δa in comparison with anisotropy of elastic modulus. It is comparable to the effect observed in polycrystalline steel and aluminum alloys.

The accumulation of hydrogen concentrations to the level of 4 ppm in a weakly bound state along the edges of specimens was revealed. The observed significant concentrations are close to limiting hydrogen concentrations for nickel-based alloys at atmospheric pressure. The critical values of hydrogen concentrations indicate degradation of mechanical properties of single-crystal specimens and their destruction by the hydrogen embrittlement mechanism.

The obtained results are important for the development of new approaches based on analysis of hydrogen and acoustic anisotropy distributions for estimating the residual life of machine parts and structures made of single-crystal materials.

Acknowledgements The research is carried out under the financial support by Russian Science Foundation, project 18-19-00413.

References

1. Toloraiya, V.N., Orekhov, N.G., Kablov, E.N.: Advanced method for single crystal casting of turbine blades for gas turbine engines and plants. Met. Sci. Heat Treat. **44**(7–8), 279–283 (2002). https://doi.org/10.1023/A:1021299619346
2. Castellano, A., Fraddosio, A., Piccioni, M.D.: Quantitative analysis of QSI and LVI damage in GFRP unidirectional composite laminates by a new ultrasonic approach. Compos. B. Eng. **151**, 106–117 (2018). https://doi.org/10.1016/j.compositesb.2018.06.003
3. Erofeev, V.I., Ilyakhinsky, A.V., Nikitina, E.A., Pakhomov, P.A., Rodyushkin, V.M.: Ultrasonic sensing method for evaluating the limit state of metal structures associated with the onset of plastic deformation. Phys. Mesomech. **23**(2), 241–245 (2020). https://doi.org/10.1134/S102995992003008X

4. Semenov, A.S., Polyanskiy, V.A., Tretyakov, D.A., Shtukin, L.V.: Effect of surface layer damage on acoustic anisotropy. J. Appl. Mech. Tech. Phys. **59**(6), 1136–1144 (2018). https://doi.org/10.1134/S0021894418060202
5. Belyaev, A.K., Polyanskiy, V.A., Semenov, A.S., Tretyakov, D.A., Yakovlev, Y.A.: Investigation of the correlation between acoustic anisotropy, damage and measures of the stress-strain state. Procedia Struct. Integrity **6**, 201–207 (2017). https://doi.org/10.1016/j.prostr.2017.11.031
6. Polyanskiy, V.A., Frolova, K., Tretyakov, D.A., Yakovlev, Y.A.: Identification of zones of local hydrogen embrittlement of metals by the acoustoelastic effect. Adv. Mater. **224**, 495–503 (2019). https://doi.org/10.1007/978-3-030-19894-7_38
7. Polyanskiy, V.A., Belyaev, A.K., Alekseeva, E.L., Polyanskiy, A.M., Tretyakov, D.A., Yakovlev, Y.A.: Phenomenon of skin effect in metals due to hydrogen absorption. Contin. Mech. Thermodyn. **31**(6), 1961–1975 (2019). https://doi.org/10.1007/s00161-019-00839-2
8. Tretyakov, D.A., Belyaev, A.K.: Surface effect of acoustic anisotropy during plastic deformation of metals. AIP Conf. Proc. **2176**(1), 020011 (2019). https://doi.org/10.1063/1.5135123
9. Brillouin, L.: Sur les tensions de radiation. Ann. Phys. **10**(4), 528–586 (1925). https://doi.org/10.1051/anphys/192510040528
10. Bridgman, P.W.: The effect of pressure on the rigidity of several metals. Proc. Amer. Acad. Arts Sci. **64**(3), 39–49 (1929). https://doi.org/10.2307/20026251
11. Bridgman, P.W.: The compression of 39 substances to 100,000 Kg/Cm. Proc. Amer. Acad. Arts Sci. **76**(3), 55–70 (1948)
12. Birch, F., Bancroft, D.: The effect of pressure on the rigidity of rocks. I. J. Geol. **46**(1), 59–87 (1938)
13. Biot, M.A.: The influence of initial stress on elastic waves. J. Appl. Phys. **11**(8), 522–530 (1940). https://doi.org/10.1063/1.1712807
14. Hughes, D.S., Kelly, J.L.: Second-order elastic deformation of solids. Phys. Rev. **92**(5), 1145–1149 (1953). https://doi.org/10.1103/PhysRev.92.1145
15. Murnaghan, F.D.: Finite deformations of an elastic solid. Am. J. Math. **59**(2), 235–260 (1937). https://doi.org/10.2307/2371405
16. Benson, R.W., Raelson, V.J.: Acoustoelasticity. Prod. Eng. **30**(29), 56–59 (1959)
17. Toupin, R.A., Bernstein, B.: Sound waves in deformed perfectly elastic materials. Acoustoelastic Effect. J. Acoust. Soc. Am. **33**(2), 216–225 (1961). https://doi.org/10.1121/1.1908623
18. Fukuoka, H., Toda, H., Naka, H.: Nondestructive residual-stress measurement in a wide-flanged rolled beam by acoustoelasticity. Exp. Mech. **23**(1), 120–128 (1983). https://doi.org/10.1007/BF02328690
19. Tokuoka, T., Saito, M.: Elastic wave propagations and acoustical birefringence in stressed crystals. J. Acoust. Soc. Am. **45**(5), 1241–1246 (1969). https://doi.org/10.1121/1.1911595
20. Okada, K.: Stress-acoustic relations for stress measurement by ultrasonic technique. J. Acoust. Soc. Jpn (e) **1**(3), 193–200 (1980). https://doi.org/10.1250/ast.1.193
21. Bateman, T., Mason, W.P., McSkimin, H.J.: Third-order elastic moduli of germanium. J. Appl. Phys. **32**(5), 928–936 (1961). https://doi.org/10.1063/1.1736135
22. McSkimin, H.J., Andreatch Jr., P.: Measurement of third-order moduli of silicon and germanium. J. Appl. Phys. **35**(11), 3312–3319 (1964). https://doi.org/10.1063/1.1713214
23. Thurston, R.N., McSkimin, H.J., Andreatch Jr., P.: Third-order elastic coefficients of quartz. J. Appl. Phys. **37**(1), 267–275 (1966). https://doi.org/10.1063/1.1707824
24. Smith, R.T., Stern, R., Stephens, R.W.B.: Third-order elastic moduli of polycrystalline metals from ultrasonic velocity measurements. J. Acoust. Soc. Am. **40**(5), 1002–1008 (1966). https://doi.org/10.1121/1.1910179
25. Smith, R.T.: Stress-induced anisotropy in solids—the acousto-elastic effect. Ultrasonics **1**(3), 135–147 (1963). https://doi.org/10.1016/0041-624X(63)90003-9
26. Crecraft, D.I.: The measurement of applied and residual stresses in metals using ultrasonic waves. J. Sound Vib. **5**(1), 173–192 (1967). https://doi.org/10.1016/0022-460X(67)90186-1
27. Hsu, N.N.: Acoustical birefringence and the use of ultrasonic waves for experimental stress analysis. Exp. Mech. **14**(5), 169–176 (1974). https://doi.org/10.1007/BF02323061

28. Egle, D.M., Bray, D.E.: Measurement of acoustoelastic and third-order elastic constants for rail steel. J. Acoust. Soc. Am. **60**(3), 741–744 (1976). https://doi.org/10.1121/1.381146
29. King, R.B., Herrmann, G., Kino, G.S.: Use of stress measurements with ultrasonics for non-destructive evaluation of the J integral. Eng. Fract. Mech. **15**(1–2), 77–86 (1981). https://doi.org/10.1016/0013-7944(81)90107-7
30. Salama, K., Ling, C.K.: The effect of stress on the temperature dependence of ultrasonic velocity. J. Appl. Phys. **51**(3), 1505–1509 (1980). https://doi.org/10.1063/1.327800
31. Erofeev, V.I., Zaznobin, V.A., Samokhvalov, R.V.: Determination of mechanical stresses in solids by an acoustic method. Acoust. Phys. **53**(5), 546–552 (2007). https://doi.org/10.1134/S1063771007050028
32. Nikitina, N.E., Kamyshev, A.V., Kazachek, S.V.: Ultrasonic testing of the stressed state of pipelines with consideration of the temperature factor. Russ. J. Nondestruct. Test. **48**(5), 272–276 (2012). https://doi.org/10.1134/S1061830912050087
33. Nikitina, N.E., Kazachek, S.V., Kulizina, O.V., Kazachek, Y.N.: Survey of plain stress condition of a perforated plate by the acoustoelasticity method. J. Mach. Manuf. Reliab. **39**(8), 554–558 (2010). https://doi.org/10.3103/S1052618810060075
34. Castellano, A., Fraddosio, A., Marzano, S., Piccioni, M.D.: Some advancements in the ultrasonic evaluation of initial stress states by the analysis of the acoustoelastic effect. Procedia Eng. **199**, 1519–1526 (2017). https://doi.org/10.1016/j.proeng.2017.09.494
35. Ivanova, Y., Partalin, T., Pashkuleva, D.: Acoustic investigations of the steel samples deformation during the tensile. Russ. J. Nondestruct. Test. **53**(1), 39–50 (2017). https://doi.org/10.1134/S1061830917010077
36. Pao, Y.H., Gamer, U.: Acoustoelastic waves in orthotropic media. J. Acoust. Soc. Am. **77**(3), 806–812 (1985). https://doi.org/10.1121/1.392384
37. Kobayashi, M.: Theoretical study of acoustoelastic effects caused by plastic anisotropy growth. Int. J. Plast. **3**(1), 1–20 (1987). https://doi.org/10.1016/0749-6419(87)90014-3
38. Kobayashi, M.: Ultrasonic nondestructive evaluation of microstructural changes of solid materials under plastic deformation-Part I. Theory. Int. J. Plast. **14**(6), 511–522 (1998). https://doi.org/10.1016/S0749-6419(98)00005-9
39. Kobayashi, M.: Ultrasonic nondestructive evaluation of microstructural changes of solid materials under plastic deformation-Part II. Exp. Simul. Int. J. Plast. **14**(6), 523–535 (1998). https://doi.org/10.1016/S0749-6419(98)00006-0
40. Tretyakov, D.A., Belyaev, A.K., Galyautdinova, A.R., Polyanskiy, V.A., Strekalovskaya, D.A.: Investigation of the corrosion process and destruction of metals by using Acoustodamage method. E3S Web of Conferences **121**, 01017 (2019). https://doi.org/10.1051/e3sconf/201912101017
41. Seeger, A., Buck, O.: Die experimentelle Ermittlung der elastischen Konstanten höherer Ordnung. Z. Naturforsch. A **15**(12), 1056–1067 (1960). https://doi.org/10.1515/zna-1960-1205
42. Hikata, A., Truell, R., Granato, A., Chick, B., Lücke, K.: Sensitivity of ultrasonic attenuation and velocity changes to plastic deformation and recovery in aluminum. J. Appl. Phys. **27**(4), 396–404 (1956). https://doi.org/10.1063/1.1722383
43. Hikata, A., Chick, B., Elbaum, C., Truell, R.: Ultrasonic attenuation and velocity data on aluminum single crystals as a function of deformation and orientation. Acta Metall. **10**(4), 423–429 (1962). https://doi.org/10.1016/0001-6160(62)90021-4
44. Einspruch, N.G., Manning, R.J.: Third-order elastic moduli of anisotropic solids. J. Appl. Phys. **35**(3), 560–567 (1964). https://doi.org/10.1063/1.1713416
45. Thurston, R.N.: Effective elastic coefficients for wave propagation in crystals under stress. J. Acoust. Soc. Am. **37**(2), 348–356 (1965). https://doi.org/10.1121/1.1909333
46. Motova, E.A., Nikitina, N.E., Tarasenko, Y.P.: Concerning the possibility of examining compressor blades according to attenuation and speed of ultrasound. J. Mach. Manuf. Reliab. **42**(4), 335–340 (2013). https://doi.org/10.3103/S1052618813040109
47. Tarasenko, Yu.P., Nikitina, N.E., Krivina, L.A., Motova, E.A.: The study of the structural and acoustic anisotropy of the material of high-pressure turbine compressor blades. J. Mach. Manuf. Reliab. **44**(2), 186–188 (2015). https://doi.org/10.3103/S1052618815020132

48. Getsov, L.B., Semenov, A.S., Tikhomirova, E.A., Rybnikov, A.I.: Thermocyclic- and static-failure criteria for single-crystal superalloys of gas-turbine blades. Mater. Technol. **48**(2), 255–260 (2014)
49. Getsov, L.B., Dobina, N.I., Rybnikov, A.I., Semenov, A.S., Staroselskii, A., Tumanov, N.V.: Thermal fatigue resistance of a monocrystalline alloy. Strength of materials **40**(5), 538–551 (2008). https://doi.org/10.1007/s11223-008-9076-1
50. Konar, J., Banerjee, N.G.: Vacuum heating hydrogen determination in aluminium and aluminium alloys. NML Techn. J. **16**(1–2), 18–19 (1974)

Chapter 3
Skin Effect of Acoustic Anisotropy and Dissolved Hydrogen in Metals

Alexander K. Belyaev, Aliya R. Galyautdinova, Vladimir A. Polyanskiy, and Dmitry A. Tretyakov

Abstract The work is devoted to study the influence of damaged surface layer on mechanical properties of metal structures. Comparative studies of acoustic anisotropy distributions and dissolved hydrogen concentrations measured according to the vacuum heating method were carried out. The correlation of distributions in rolled steel and aluminum specimens after elasto-plastic and fatigue destruction was revealed. The obtained results indicate the possibility of using the acoustoelasticity method for detecting localized plastic deformations, surface microcracking, and zones with increased concentrations of dissolved hydrogen in metal. It can be used to develop new nondestructive ultrasonic approaches in technical diagnostics of metal structures.

Keywords Acoustic birefringence · Dissolved hydrogen · Elasto-plastic deformation · Nondestructive ultrasonic testing · Vacuum heating method

3.1 Inroduction

The acoustoelasticity method is an ultrasonic nondestructive testing method used to estimate mechanical stresses in solids [1–7]. It is based on phenomenon of acoustic birefringence observed in elastically stressed materials with a crystalline structure. This phenomenon is also called the acoustoelastic effect. It consists in splitting elastic shear waves into components V_1, V_2 polarized along axes of principal stresses σ_1, σ_2

A. K. Belyaev · A. R. Galyautdinova · V. A. Polyanskiy · D. A. Tretyakov (✉)
Peter the Great St. Petersburg Polytechnic University (SPbPU), St. Petersburg, Russia
e-mail: tretyakov_da@spbstu.ru

A. K. Belyaev
e-mail: vice.ipme@gmail.com

A. R. Galyautdinova
e-mail: aliyagalyautdinova@gmail.com

V. A. Polyanskiy
e-mail: vapol@mail.ru

V. A. Polyanskiy and A. K. Belyaev (eds.), *Advances in Hydrogen Embrittlement Study*, Advanced Structured Materials 143, https://doi.org/10.1007/978-3-030-66948-5_3

in case of plane stress state. The acoustoelastic effect was theoretically predicted in 1930s and 1940s by Biot [8], Birch [9], Lazarus [10], and others.

The modern technology of acoustoelastic measurements is based on the theory of wave propagation in a nonlinearly elastic stressed medium. It was obtained after the research was carried out by Hughes and Kelly [11], Toupin and Bernstein [12], Truesdell [13], Tokuoka and Saito [14], Iwashimizu and Kubomura [15], Okada [16, 17], Clark [18] and others. Generalization of results obtained for more than fifty years was carried out by Fukuoka in [19, 20].

The essence of results proposed in [11–17] is contained in relation (3.1) obtained by Hirao and Pao in [21, 22] for a weakly anisotropic material:

$$\frac{V_1 - V_2}{V_0} = \frac{C_{55} - C_{44}}{2\mu} + \frac{4\mu + n}{8\mu^2}(\sigma_1 - \sigma_2) + \alpha_1(\varepsilon_1^p - \varepsilon_2^p), \tag{3.1}$$

where $\Delta a = (V_1 - V_2)/V_0$ is non-dimensional parameter of total acoustic anisotropy of material, μ is the second-order Lame constant, n is the third-order Murnaghan constant [23], C_{44} and C_{55} are elastic constants in Voigt's notation, α_1 is the experimentally determined constant according to Hirao and Pao [21, 22], σ_1, σ_2 are the principal biaxial stresses, ε_1^p, ε_2^p are the principal plastic deformations. Nondestructive testing methods based on estimating value of acoustic birefringence widely use the acoustic anisotropy parameter Δa.

Classical acoustoelasticity is limited by the use of first two components in relation (3.1). The first component $\alpha_0 = (C_{55} - C_{44})/2\mu$ is related with intrinsic anisotropy caused by microstructure and anisotropic texture of material. The second component $((4\mu + n)/8\mu^2)(\sigma_1 - \sigma_2)$ is related with the acoustoelastic effect in case of an elastically stressed medium. The experimental results of linear acoustoelasticity were obtained by Benson and Raelson [24], Smith [25], Crecraft [26, 27], Hsu [28], Papadakis [29], Blinka and Sachse [30], Egle and Bray [31, 32], Kino [33], King [34], Janssen [35] and others.

Hirao and Pao [21, 22] proposed to expand application of the acoustoelasticity method [24] by including component $\alpha_1(\varepsilon_1^p - \varepsilon_2^p)$ linearly related to plastic deformations (3.1). It was experimentally verified in the case of four-point bending of annealed prismatic beams at plastic deformations of 1% order [21]. The study of small plastic deformations according to relation (3.1) was also carried out in [36].

More than 30 years have passed since the appearance of [21, 22] results. Recent studies indicate a significant influence of damage accumulated during monotonic [37] and cyclic loading [38] on acoustic anisotropy of industrial structures. In addition, influence of hydrogen embrittlement on ultrasonic measurements was found during standard hydrogen-induced corrosion tests (HIC) [39].

At the same time, uneven accumulation of plastic deformation and damage is usually observed. The periodic structure of plastic deformation waves called "chessboard" is observed in polycrystalline materials [40–42]. Microcracks localized in a thin surface layer have a significant influence on the degradation of mechanical properties of metal [43, 44]. Therefore, the use of relation (3.1) obtained for specially prepared isotropic specimens may be incorrect for real structures.

The aim of this work is to carry out comparative studies of acoustic anisotropy and high-precision measurements of hydrogen concentrations in rolled specimens from steel and aluminum alloys.

3.2 Experiments

3.2.1 Methods

Three stages of experimental research were carried out. At the first stage, metal specimens cut from rolled sheet were subjected to uniaxial loading. Thus, the Hirao–Pao relationship (3.1) was tested in a wide range of loads and deformations.

At the second stage, ultrasonic measurements were carried out using the Benson–Raelson approach [24] with calculation of total acoustic anisotropy Δa according to relation (3.2):

$$\Delta a = \frac{V_1 - V_2}{(V_1 + V_2)/2}, \tag{3.2}$$

where denominator $(V_1 + V_2)/2$ is used instead of shear wave velocity V_0 for an unstressed isotropic material.

Ultrasonic measurements were carried out using a standard sensor with 5 MHz acoustic signal frequency. Wave packets were emitted and received by 12 × 12 mm piezoelectric transducers. The ultrasonic sensor was controlled by an acoustic anisotropy analyzer [5]. This device allowed one to obtain average time delays t_1, t_2 between multiple reflected pulses. The velocities V_1, V_2 were obtained after direct micrometric measurement of specimen thickness h. Shear waves were polarized along and perpendicular to direction of uniaxial loading.

At the third stage, hydrogen concentrations were measured by the vacuum heating method [45, 46]. It is based on use of hydrogen diffusion during heating of analyzed metal. The specimen extraction curves contain dependence of extracted hydrogen flux on heating time in vacuum.

Dissolved hydrogen measurements were carried out using high-precision industrial analyzer. Small prismatic specimens with 8 mm-size were tested. The experimental technique for determining hydrogen concentrations is described in [47].

3.2.2 Acoustoelastic Effect in Case of Uniaxial Elasto-Plastic Deformation

The total acoustic anisotropy Δa and velocities of ultrasonic shear waves V_1, V_2 in the case of uniaxial elasto-plastic tension were investigated. Cold-rolled 500 × 70 × 15 mm specimens from AMts aluminum alloy were tested. Uniaxial rigid loading was carried out with 5 mm step on INSTRON-8806 hydraulic tensile testing machine.

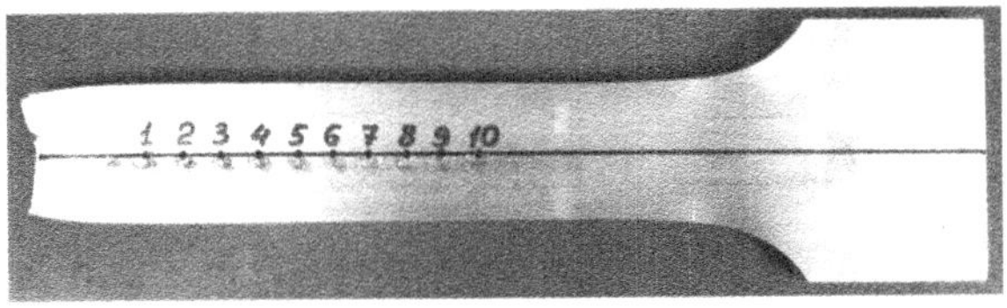

Fig. 3.1 Aluminum specimen after elasto-plastic destruction

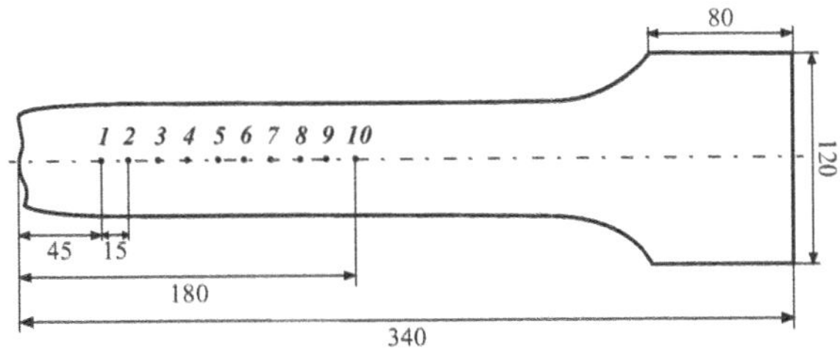

Fig. 3.2 Schematic view of measurement points location

One of the specimens after destruction and unloading is shown in Fig. 3.1. It was subjected to 19 loading stages. Acoustic measurements were carried out at 10 points located along the specimen axis (in Fig. 3.2). The total axial deformation ε up to specimen destruction was equal to 26.70%. The initial distribution of acoustic anisotropy Δa obtained along specimen axis turned out to be close to uniform and was equal to $a_0 = 0.50 \pm 0.03\%$.

Measurement points No. 1, 4, 7, and 10 were located at 45, 90, 135- and 180-mm distance from destruction zone. The results of acoustic anisotropy measurements at each loading stage are shown in Fig. 3.3.

3.2.3 Comparative Studies of Acoustic Anisotropy and Hydrogen Concentrations

Comparative studies of acoustic anisotropy distributions Δa, % and hydrogen concentrations CH, ppm were carried out at $n = 10$ points located along specimen axis (see Fig. 3.2). The hydrogen concentrations were investigated including surface layer and inside the aluminum specimen at 2 mm depth from its surface. The obtained results are shown in Figs. 3.4 and 3.5.

Also, studies on rolled COR-TEN steel specimen (in Fig. 3.6) were carried out. The specimen was subjected to 380 kN cyclic load and destroyed after $n = 5 \times 10^6$ loading cycles (in Fig. 3.7).

Similar studies of acoustic anisotropy Δa, % and dissolved hydrogen concentrations CH, ppm were carried out at $n = 10$ points located along steel specimen axis. The results obtained near fatigue destruction zone are shown in Figs. 3.8 and 3.9.

Fig. 3.3 Dependence of acoustic anisotropy Δa on elasto-plastic deformations ε

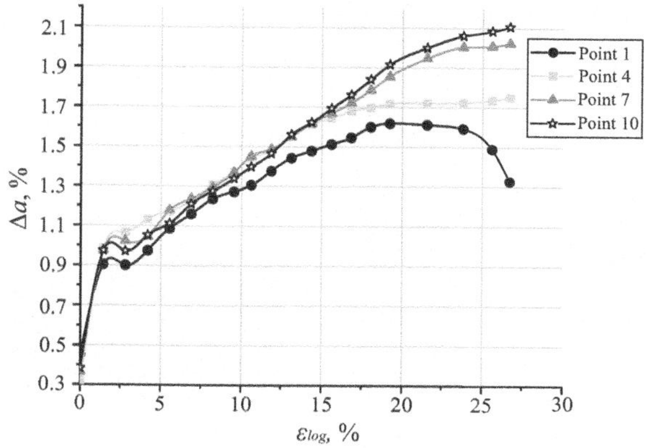

Fig. 3.4 Acoustic anisotropy Δa, % at n points after elasto-plastic destruction of aluminum specimen

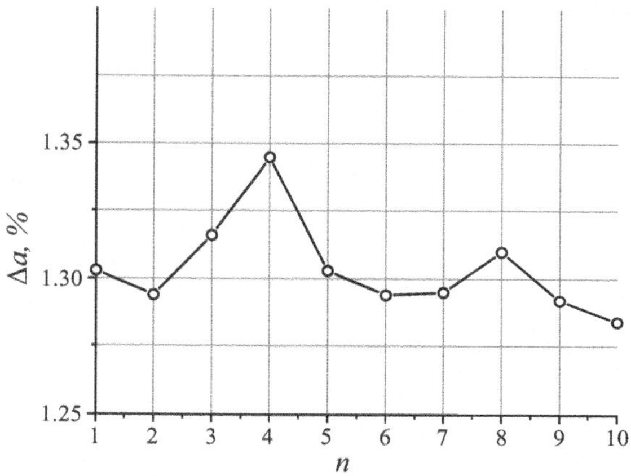

Fig. 3.5 Hydrogen concentration CH, ppm at n points before and after removing surface layer of aluminum specimen

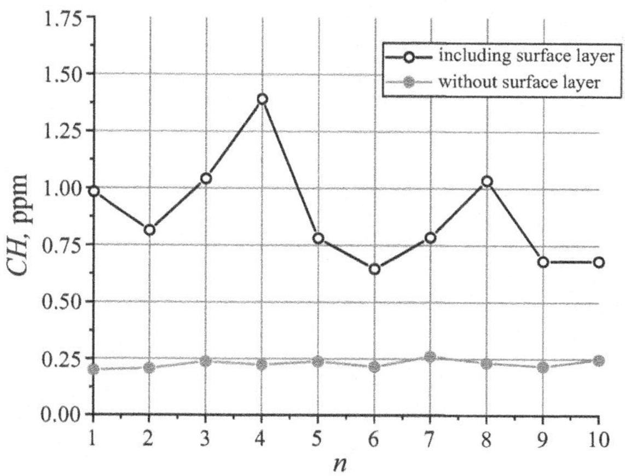

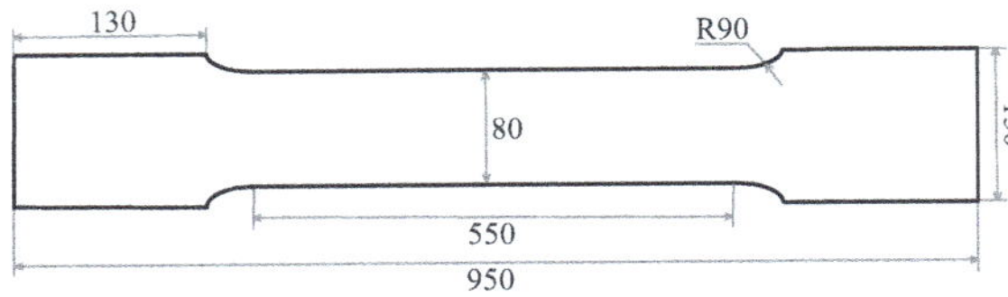

Fig. 3.6 Schematic view of steel specimen

Fig. 3.7 Steel specimen after fatigue destruction

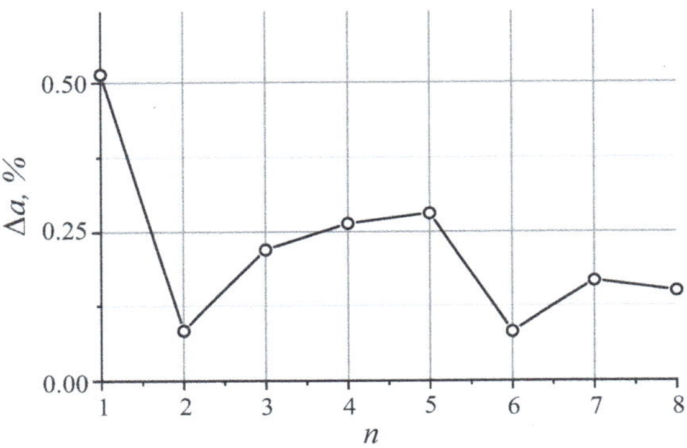

Fig. 3.8 Acoustic anisotropy Δa, % at n points after fatigue destruction of steel specimen

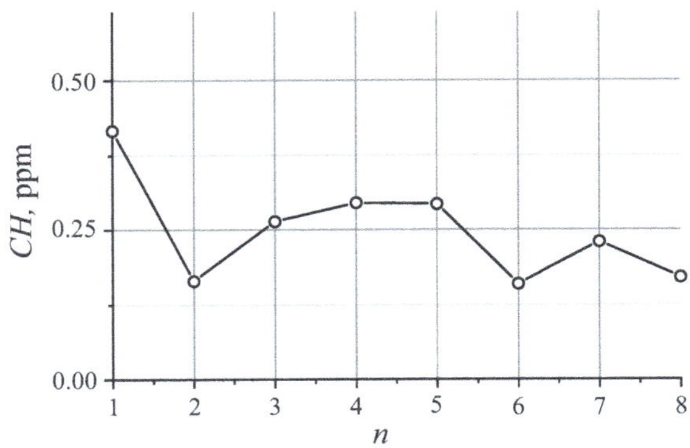

Fig. 3.9 Hydrogen concentration CH, ppm at n points after fatigue destruction of steel specimen

3.3 Discussion

3.3.1 *Nonlinear Dependence of Acoustic Anisotropy During Elasto-Plastic Deformation*

Several important conclusions follow from results presented in Fig. 3.3. The contribution of elastic deformations to the value of acoustic anisotropy averaged 0.54%. It is comparable to initial acoustic anisotropy due to influence of anisotropic texture. Thus, the contribution of component a_0 must be studied during acoustoelastic measurements in cold-rolled structures.

The dependence of acoustic anisotropy is nonlinear with increasing deformations (see Fig. 3.3). All experimental curves have non-monotonic region at deformations equal to $\varepsilon = 1.42 \div 2.82\%$. It is observed with the beginning of plastic flow process.

Two other non-monotonic regions are observed for points No. 1 and 4 after deformations equal to $\varepsilon = 19.20\%$. Significant changes in acoustic anisotropy value are related with the specimen surface curvature due to plastic neck formation. Semenov [37, 43] obtained relation (3.3) for a small deviation α angle from the surface normal. In this case, acoustic anisotropy Δa is calculated as

$$\Delta a = \alpha^2 \frac{9(\lambda + \mu)}{(4(\lambda + \mu) - r)} \frac{r}{\mu}, \tag{3.3}$$

where λ, μ are the Lame constants, $r = 4\mu^2/(H' + 3\mu)$, H' is inclination angle of stress–strain diagram [37, 43].

The actual dependence of acoustic anisotropy on plastic deformations has a nonlinear nonmonotonic character (see Fig. 3.3) and cannot be described by linear dependence $\alpha_1(\varepsilon_1^p - \varepsilon_2^p)$ from relation (3.1). It differs from previously published results obtained on basis of the Murnaghan's theory of finite deformations [23]. Alternative mathematical models [37] should be used to describe propagation of shear waves at large plastic deformations of material.

3.3.2 *Skin Effect of Acoustic Anisotropy and Hydrogen Concentrations*

The formation of metal defects in atmosphere not subject to drying or filling with inert gases is related to an increase in hydrogen concentrations. Hydrogen is an indicator of structural changes in metals such as cracking, loosening, pore formation, and others. It also leads to formation of new defects under external loading.

The diffusion of hydrogen has a great influence on processes of corrosion, cracking, and brittle fracture of metals. The mechanism of hydrogen embrittlement was explained by Gorsky [48]. He found that deformations of crystalline matrix caused by mechanical stresses have an influence on hydrogen diffusion in solids.

Comparative studies of hydrogen concentrations and acoustic anisotropy distributions revealed their correlation both on aluminum (in Figs. 3.4 and 3.5) and steel specimens (in Figs. 3.8 and 3.9). This effect cannot be due to residual stresses or influence of crystal structure. The observed macroscopic distributions significantly exceed the effects related with plastic flow, such as formation of Luders bands or the Portevin-Le Chatelier effect [49]. It is similar to the phenomenon of propagation of plastic flow autowaves discovered by Zuev [50].

At the same time, the qualitative coincidence of ultrasonic and hydrogen results can be due to influence of microcracks leading to material loosening under monotonic or cyclic load. The uniform distribution of hydrogen inside the specimen (in Fig. 3.5) indicates the surface effect of hydrogen accumulation [44]. It also indirectly indicates the surface effect of acoustic anisotropy predicted earlier in [43].

3.4 Conclusions

Experimental studies have revealed nonlinear non-monotonic character of the acoustic anisotropy dependence on elasto-plastic deformations for orthotropic specimens. This effect is related with the beginning of different stages of deformation process. It cannot be described using generally accepted acoustoelasticity relations postulating linear dependence on plastic deformations. Thus, the influence of large plastic deformations on shear wave velocities should be described using alternative models not based on Murnaghan's theory of finite deformations.

The revealed correlation between acoustic anisotropy distributions and dissolved hydrogen concentrations can be explained only by formation of microcracks systems localized in a thin surface layer of metal. The results of comparative ultrasonic and hydrogen studies indicate the possibility of detecting metal saturation with hydrogen using acoustic anisotropy measurements.

The obtained results allow one to apply the acoustoelasticity method to nondestructive testing of plastic deformations and microcracking of metal structures at any stage of their operation.

Acknowledgements The research is carried out under the financial support by Russian Science Foundation, project 18-19-00413.

References

1. Mishakin, V.V., Serebryany, V.N.: Application of the acoustoelastic effect in the precise evaluation of the plastic strain value. Acoustics letters. **17**(7), 123–128 (1994)
2. Santos Jr., A.A., Bray, D.E.: Comparison of acoustoelastic methods to evaluate stresses in steel plates and bars. J. Press. Vessel Technol. **124**(3), 354–358 (2002). https://doi.org/10.1115/1.1484114

3. Erofeev, V.I., Zaznobin, V.A., Samokhvalov, R.V.: Determination of mechanical stresses in solids by an acoustic method. Acoust. Phys. **53**(5), 546–552 (2007). https://doi.org/10.1134/S1063771007050028
4. Kobayashi, H., Vanderby, R.: Acoustoelastic analysis of reflected waves in nearly incompressible, hyper-elastic materials: forward and inverse problems. J. Acoust. Soc. Am. **121**(62), 879–887 (2007). https://doi.org/10.1121/1.2427112
5. Kamyshev, A.V., Nikitina, N.E., Smirnov, V.A.: Measurement of the residual stresses in the treads of railway wheels by the acoustoelasticity method. Russ. J. Nondestruct. Test. **46**(3), 189–193 (2010). https://doi.org/10.1134/S106183091003006X
6. Castellano, A., Foti, P., Fraddosio, A., Marzano, S., Paparella, F., Piccioni, M.D.: Monitoring applied and residual stress in materials and structures by non-destructive acoustoelastic techniques. IEEE **1–5** (2016). https://doi.org/10.1109/EESMS.2016.7504830
7. Ivanova, Y., Partalin, T., Pashkuleva, D.: Acoustic investigations of the steel samples deformation during the tensile. Russ. J. Nondestruct. Test. **53**(1), 39–50 (2017). https://doi.org/10.1134/S1061830917010077
8. Biot, M.A.: XLIII. Non-linear theory of elasticity and the linearized case for a body under initial stress. Philos. Mag. **217**(7), 468–489 (1939). https://doi.org/10.1080/14786443908562246
9. Birch, F., Bancroft, D.: The effect of pressure on the rigidity of rocks. I. J. Geol. **46**(1), 59–87 (1938)
10. Lazarus, D.: The variation of the adiabatic elastic constants of KCl, NaCl, CuZn, Cu, and Al with pressure to 10,000 bars. Phys. Rev. **76**(4), 545–553 (1949). https://doi.org/10.1103/PhysRev.76.545
11. Hughes, D.S., Kelly, J.L.: Second-order elastic deformation of solids. Phys. Rev. **92**(5), 1145–1149 (1953). https://doi.org/10.1103/PhysRev.92.1145
12. Toupin, R.A., Bernstein, B.: Sound waves in deformed perfectly elastic materials. Acoustoelastic effect. J. Acoust. Soc. Am. **33**(2), 216–225 (1961). https://doi.org/10.1121/1.1908623
13. Truesdell, C.: General and exact theory of waves in finite elastic strain. Arch. Ration Mech. Anal. **8**(1), 263–296 (1961). https://doi.org/10.1007/BF00277444
14. Tokuoka, T., Saito, M.: Elastic wave propagations and acoustical birefringence in stressed crystals. J. Acoust. Soc. Am. **45**(5), 1241–1246 (1969). https://doi.org/10.1121/1.1911595
15. Iwashimizu, Y., Kubomura, K.: Stress-induced rotation of polarization directions of elastic waves in slightly anisotropic materials. Int. J. Solids Struct. **9**(1), 99–114 (1973). https://doi.org/10.1016/0020-7683(73)90035-8
16. Okada, K.: Stress-acoustic relations for stress measurement by ultrasonic technique. J. Acoust. Soc. Jpn (e) **1**(3), 193–200 (1980). https://doi.org/10.1250/ast.1.193
17. Okada, K.: Acoustoelastic determination of stress in slightly orthotropic materials. Exp. Mech. **21**(12), 461–466 (1981). https://doi.org/10.1007/BF02327418
18. Clark, A.V., Mignogna, R.B.: A comparison of two theories of acoustoelasticity. Ultrasonics **21**(5), 217–225 (1983). https://doi.org/10.1016/0041-624X(83)90045-8
19. Fukuoka, H., Toda, H., Naka, H.: Nondestructive residual-stress measurement in a wide-flanged rolled beam by acoustoelasticity. Exp. Mech. **23**(1), 120–128 (1983). https://doi.org/10.1007/BF02328690
20. Fukuoka, H.: Ultrasonic Measurement of Residual Stress. Solid mechanics research for quantitative non-destructive evaluation. Springer, Dordrecht, 275–299 (1987). https://doi.org/10.1007/978-94-009-3523-5_17
21. Hirao, M., Pao, Y.H.: Dependence of acoustoelastic birefringence on plastic strains in a beam. J. Acoust. Soc. Am. **77**(5), 1659–1664 (1985). https://doi.org/10.1121/1.391964
22. Pao, Y.H.: Theory of acoustoelasticity and acoustoplasticity. Solid mechanics research for quantitative non-destructive evaluation. Springer, Dordrecht, 257–273 (1987). https://doi.org/10.1007/978-94-009-3523-5_16
23. Murnaghan, F.D.: Finite deformations of an elastic solid. Am. J. Math. **59**(2), 235–260 (1937). https://doi.org/10.2307/2371405
24. Benson, R.W., Raelson, V.J.: Acoustoelasticity. Prod. Eng. **30**(29), 56–59 (1959)

25. Smith, R.T.: Stress-induced anisotropy in solids—the acousto-elastic effect. Ultrasonics **1**(3), 135–147 (1963). https://doi.org/10.1016/0041-624X(63)90003-9
26. Crecraft, D.I.: The use of ultrasonics in stress analysis. Strain **1**(4), 4–8 (1965). https://doi.org/10.1111/j.1475-1305.1965.tb00064.x
27. Crecraft, D.I.: The measurement of applied and residual stresses in metals using ultrasonic waves. J. Sound Vib. **5**(1), 173–192 (1967). https://doi.org/10.1016/0022-460X(67)90186-1
28. Hsu, N.N.: Acoustical birefringence and the use of ultrasonic waves for experimental stress analysis. Exp. Mech. **14**(5), 169–176 (1974). https://doi.org/10.1007/BF02323061
29. Papadakis, E.P.: Ultrasonic phase velocity by the pulse-echo-overlap method incorporating diffraction phase corrections. J. Acoust. Soc. Am. **42**(5), 1045–1051 (1967). https://doi.org/10.1121/1.1910688
30. Blinka, J., Sachse, W.: Application of ultrasonic-pulse-spectroscopy measurements to experimental stress analysis. Exp. Mech. **16**(12), 448–453 (1976). https://doi.org/10.1007/BF02324101
31. Egle, D.M., Bray, D.E.: Measurement of acoustoelastic and third-order elastic constants for rail steel. J. Acoust. Soc. Am. **60**(3), 741–744 (1976). https://doi.org/10.1121/1.381146
32. Bray, D.E., Egle, D.M.: Ultrasonic studies of anisotropy in cold-worked layer of used rail. Metal Science **15**(11–12), 574–582 (1981). https://doi.org/10.1179/msc.1981.15.11-12.574
33. Kino, G.S., Hunter, J.B., Johnson, G.C., Selfridge, A.R., Barnett, D.M., Hermann, G., Steele, C.R.: Acoustoelastic imaging of stress fields. J. Appl. Phys. **50**(4), 2607–2613 (1979). https://doi.org/10.1063/1.326268
34. King, R.B., Herrmann, G., Kino, G.S.: Use of stress measurements with ultrasonics for non-destructive evaluation of the J integral. Eng. Fract. Mech. **15**(1–2), 77–86 (1981). https://doi.org/10.1016/0013-7944(81)90107-7
35. Janssen, M.: Evaluation of an applied plane-stress tensor distribution using ultrasonic shear waves. Exp. Mech. **28**(3), 226–231 (1988). https://doi.org/10.1007/BF02329015
36. Kobayashi, M.: Theoretical study of acoustoelastic effects caused by plastic anisotropy growth. Int. J. Plast. **3**(1), 1–20 (1987). https://doi.org/10.1016/0749-6419(87)90014-3
37. Belyaev, A.K., Lobachev, A.M., Modestov, V.S., Pivkov, A.V., Polyanskiy, V.A., Semenov, A.S., Tretyakov, D.A., Shtukin, L.V.: Estimating the plastic strain with the use of acoustic anisotropy. Mech. Solids **51**(5), 606–611 (2016). https://doi.org/10.3103/S0025654416050149
38. Tretyakov, D., Belyaev, A., Polyanskiy, V., Stepanov, A., Yakovlev, Yu.: Correlation of acoustoelasticity with hydrogen saturation during destruction. E3S Web of Conferences. **121**, 01016-1 – 01016-5 (2019). https://doi.org/10.1051/e3sconf/201912101016
39. Tretyakov, D.A., Belyaev, A.K., Galyautdinova, A.R., Polyanskiy, V.A., Strekalovskaya, D.A.: Investigation of the corrosion process and destruction of metals by using Acoustodamage method. E3S Web of Conferences **121**, 01017-1 – 01017-4 (2019). https://doi.org/10.1051/e3sconf/201912101017
40. Panin, V.E.: Foundations of Physical Mesomechanics. Phys. Mesomech. **1**(1), 5–22 (1998)
41. Panin, V.E., Panin, A.V., Moiseenko, D.D.: Physical Mesomechanics of a deformed solid as a multilevel system. II. Chessboard-Like Mesoeffect of the Interface in Heterogeneous Media in External Fields. Phys. Mesomech. **10**(1-2), 5–14 (2007). https://doi.org/10.1016/j.physme.2007.06.001
42. Polyanskiy, V.A., Belyaev, A.K., Grishchenko, A.I., Lobachev, A.M., Modestov, V.S., Pivkov, A.V., Tretyakov, D.A., Shtukin, L.V., Semenov, A.S., Yakovlev, Y.A.: Finite element simulation of chessboard strain localization in view of statistical spreads in polycrystal grain parameters. Phys. Mesomech. **22**(3), 188–194 (2019). https://doi.org/10.1134/S1029959919030032
43. Semenov, A.S., Polyanskiy, V.A., Tretyakov, D.A., Shtukin, L.V.: Effect of surface layer damage on acoustic anisotropy. J. Appl. Mech. Tech. Phys. **59**(6), 1136–1144 (2018). https://doi.org/10.1134/S0021894418060202
44. Polyanskiy, V.A., Belyaev, A.K., Alekseeva, E.L., Polyanskiy, A.M., Tretyakov, D.A., Yakovlev, Y.A.: Phenomenon of skin effect in metals due to hydrogen absorption. Contin. Mech. Thermodyn. **31**(6), 1961–1975 (2019). https://doi.org/10.1007/s00161-019-00839-2

45. Konar, J., Banerjee, N.G.: Vacuum heating hydrogen determination in aluminium and aluminium alloys. ML Techn. J. **16**(1–2), 18–19 (1974)
46. Sayi, Y.S., Ramakumar, K.L., Prasad, R., Yadav, C.S., Shankaran, P.S., Chhapru, G.C., Jain, H.C.: Determination of H2 and D2 content in metals and alloys using hot vacuum extraction. J. Radioanal. Nucl. Chem. **230**(1–2), 5–9 (1998). https://doi.org/10.1007/BF02387438
47. Andronov, D.Y., Arseniev, D.G., Polyanskiy, A.M., Polyanskiy, V.A., Yakovlev, Y.A.: Application of multichannel diffusion model to analysis of hydrogen measurements in solid. Int. J. Hydrog. Energy **42**(1), 699–710 (2017). https://doi.org/10.1016/j.ijhydene.2016.10.126
48. Gorsky, W.S.: Theorie der ordnungsprozesse und der diffusion in mischkristallen von CuAu. Sow. Phys. **8**, 443–456 (1935)
49. Le Chatelier, A.: Influence du temps et de latemperature sur les essais au choc. Rev. Met. Paris **6**(8), 914–917 (1909). https://doi.org/10.1051/metal/190906080914
50. Zuev, L.B., Danilov, V.I., Gorbatenko, V.V.: Autowaves of localized plastic deformation. Techn. Phys. **40**(5), 456–462 (1995)

Chapter 4
Thermal Desorption Spectra of Samples Charged by Hydrogen in Electrolyte

Anastasia A. Chevrychkina, Vladimir A. Polyanskiy, and Evgenii A. Varshavchik

Abstract In the framework of the diffusion equation, we discuss the influence of hydrogen distribution in the sample on thermal desorption spectroscopy. We show that non-uniform hydrogen distribution after precharging may significantly affect the Choo-Lee plot. We demonstrate that taking into account influence of hydrogen distribution is important for understanding experimental results. Good comparison of experimental and modeling data is also shown.

Keywords Hydrogen diffusion · Choo-Lee plot · Thermal Desorption Spectra

4.1 Introduction

Absorbed hydrogen affects the mechanical behavior of metal. This is the reason that an analysis of hydrogen transport and the distribution of hydrogen atoms in metals is important in metallurgy. Hydrogen permeability through metal was discovered as early as 1864 [1], and the study of hydrogen permeability is still of interest for many engineering applications, including in the field of energy: hydrogen, nuclear, thermonuclear, and thermal energy. Hydrogen permeability of metals is investigated by many researchers [2–4]. Another problem associated with the interaction of hydrogen and metals is hydrogen embrittlement. Hydrogen embrittlement is responsible for

A. A. Chevrychkina (✉) · V. A. Polyanskiy
Institute for Problems in Mechanical Engineering RAS, V.O., Bolshoy pr., 61, St. Petersburg 199178, Russia
e-mail: Anastasiia.Che@gmail.com

V. A. Polyanskiy
e-mail: vapol@mail.ru

E. A. Varshavchik
Peter the Great Saint-Petersburg Polytechnic University, Polytekhnicheskaya, 29., St. Petersburg 195259, Russia
e-mail: varshavchik.ea@edu.spbstu.ru

V. A. Polyanskiy and A. K. Belyaev (eds.), *Advances in Hydrogen Embrittlement Study*, Advanced Structured Materials 143, https://doi.org/10.1007/978-3-030-66948-5_4

premature and/or delayed rupture of materials and has become a concern in terms of the durability of materials. The hydrogen embrittlement phenomenon was first described by Johnson in 1875 [5]. Many papers are devoted to hydrogen embrittlement [6, 7].

It is assumed that the effect of hydrogen on the properties of metals is caused by absorbed hydrogen on traps such as dislocations, impurities, and internal interfaces, voids, micro-cracks, and other defects. Now the most developed approaches for the study of diffusion in solid are the approaches based on Fick's laws and statistical thermodynamics, which was proposed by Kirchheim in work [8]. The influence of traps is taken into account in both approaches.

McNabb and Foster [9] were the first to suggest separating flux of the lattice hydrogen concentration and the trapped hydrogen concentration in 1963.

$$\begin{cases} \dfrac{\partial C_L}{\partial t} + N_x \dfrac{\partial \theta_x}{\partial t} = D_L \dfrac{\partial^2 C_L}{\partial x^2}, \\ \\ \dfrac{\partial \theta_x}{\partial t} = kC_L(1 - \theta_x) - p\theta_x, \end{cases}$$

where C_L and D_L are hydrogen concentration and diffusivity in normal lattice, respectively; θ_x and N_x are occupancy and density of trap sites, respectively; k is the probability of hydrogen jumping from normal lattice site to trap site; and p is the probability of hydrogen releasing from trap site to normal lattice site.

Oriane showed the standard solutions of the diffusion equations with effective diffusion coefficient can be used to analyze experimental results in the case of assumption of local equilibrium for a restricted domain of degree of trap coverage in 1970 [10].

In 1979, Johnson brought a refinement on the limited capacity of traps, and proposed a clear dependence of the trap concentration on the lattice concentration [11]. Johnson model is the special case of the more general model McNabb and Foster.

Further development of the approach was made by Leblond and Dubois in 1983 [12]. They suggested introducing a new type of irreversible trap and analyzing the combined effects of reversible and irreversible traps. In 1989, the generalized model of hydrogen transport was developed by Turnbull which incorporated the combined effects of reversible and irreversible traps with varying degrees of occupancy [13]. Equations are described by

$$\begin{cases} \dfrac{\partial C}{\partial t} = D\dfrac{\partial^2 C}{\partial x^2} - N_r \dfrac{\partial \theta_r}{\partial t} - N_i \dfrac{\partial \theta_i}{\partial t}, \\ \\ \dfrac{\partial \theta_r}{\partial t} = k_r C(1 - \theta_r) - p\theta_r, \\ \\ \dfrac{\partial \theta_i}{\partial t} = k_i C(1 - \theta_i), \end{cases}$$

where subscripts r and i represent reversible and irreversible traps, respectively.

Another approach based on Fick's laws is the model of hydrogen multichannel diffusion [14]. Multiple channels associated with trap sites are assumed.

The generally accepted methods of activation energies' experimental estimation is thermal desorption mass spectroscopy (TDS) [15] and its modifications [16] and experimental layouts for measuring hydrogen sorption [17]. This method generally includes a hydrogen pre-charged material, then measuring its desorption flux in vacuum or in a constant inert gas flow during a constant heating rate. The determination of activation energy is based on the famous Kissinger equation (4.1) [18].

$$\frac{\partial \ln \frac{v_T}{T_p^2}}{\partial \frac{1}{T_p}} = \frac{U}{R}, \tag{4.1}$$

where v_T is a heating rate, T_p is a peak temperature, U is an activations energy, and R is the gas constant.

It is considered that the peaks of TDS spectra are attributed to the corresponding energy [5, 10, 13, 19]. Peak temperatures are measured at few heating rates, an activation energy of hydrogen is determined as a slope of the Choo-Lee plot for each traps' group [20].

Basically, in an analysis, the hydrogen distribution in the sample is assumed to be uniform after hydrogen precharging. The simulation of the ordinary diffusion equation with one activation energy gives a desorption curve with 2 peaks at an initial non-uniform hydrogen concentration as shown in the works [15, 21, 22]. It is noted that non-uniform surface hydrogen causes a new peak to appear at low temperatures. However, the effect of the non-uniform distribution of hydrogen on TDS curves needs more clarity.

The present article is focused on the influence of explicit non-uniform hydrogen concentration on TDS. First, the diffusion equation solution is considered taking into account the non-uniform initial data in the case of 1D Cartesian coordinates. It is shown that the dependence $\ln(v_T/T_p^2)$ versus $1/T_p$ on Choo-Lee plot in the case of a non-uniform initial hydrogen distribution has a nonlinear form. Numerical calculation according to Fick's laws with the non-uniform initial conditions gives excellent agreement with the experimental data. It is shown in the example of TDS data for steel from paper [23]. The calculation of the diffusion equation is carried by the finite difference method in the case of axisymmetric cylindrical coordinates.

4.2 Dependence of Peak Temperature on Flux and Heating Rate

An ordinary diffusion equation in Cartesian coordinates with a specific piece linear initial condition is reviewed. The system of diffusion equations for a sample half with a boundary and initial condition follows:

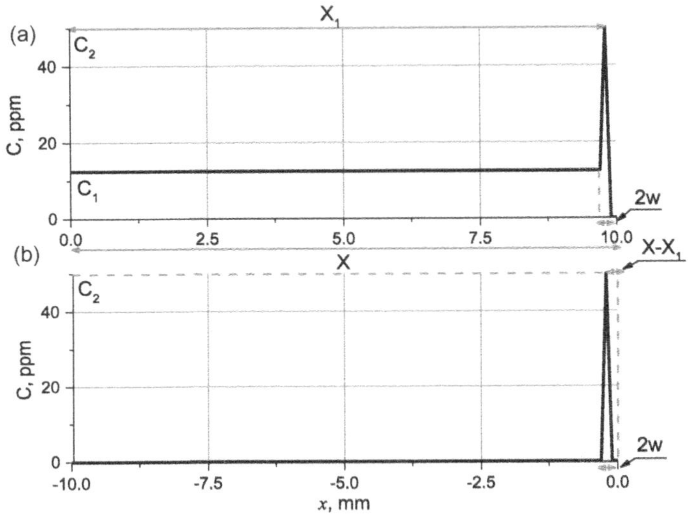

Fig. 4.1 The schematic function $C_{in}(x)$ of initial condition for the system of equations **a** (4.2) and **b** (4.5)

$$\begin{cases} \frac{1}{D(t)}\frac{\partial C}{\partial t} = \frac{\partial^2 C}{\partial x^2}, \\ C(t,x)|_{x=X} = 0, \\ \frac{\partial C(t,x)}{\partial x}|_{x=0} = 0, \\ C(0,x) = C_{in}(x), \end{cases} \tag{4.2}$$

where C is a hydrogen concentration, $D(t) = D_0 \exp\left(\frac{-U}{RT(t)}\right)$ is a diffusion coefficient, D_0 is a pre-exponential coefficient, U is an activations energy, and R is the gas constant and $T(t) = v_T t + T_0$ is a linear function of temperature dependent on time, v_T is a constant heating rate. $C_{in}(x)$ is a piecewise linear function of the initial condition, it is shown schematically on Fig. 4.1a. The initial condition corresponds to the situation, which can be as a result after hydrogen pre-charged, namely, a constant initial concentration is in the sample middle and a high hydrogen concentration is at the sample edge. Artificial saturation that is carried out in solutions leads to the fact that hydrogen accumulates at the boundary of the sample and does not pass inside. The possibility of such a hydrogen distribution in a sample after saturation is mentioned in the experimental work [23]. It is supposed in paper [21] that hydrogen gradient results from the diffusion of the adsorbate in the bulk.

Analytical solution for equation (4.2) can be written using the Fourier method. Hydrogen flux at edges is the most useful for analysis TDS, and the flow formula in the case of one-dimensional Cartesian coordinates is shown below:

$$q_H = -D(t)\frac{\partial C}{\partial x}|_{x=X} = D(t)\sum_{n=0}^{\infty} a_n T_n(t), \tag{4.3}$$

where $\lambda_n = \left(\frac{1}{2}+n\right)\frac{\pi}{X}$, $I(T) = \int_0^t \exp\left(\frac{-U}{kT(s)}\right)ds$, $T_n(t) = \exp\left(-D_0\lambda_n^2 I(t)\right)$ and coefficient a_n is given by

$$a_n=(-1)^n\frac{2\sin(w\lambda_n/2)}{w\lambda_n X}\left[(C_2-C_1)\sin\left(\lambda_n(X_1-\frac{w}{2})\right)-C_2\sin\left(\lambda_n(X_1+\frac{w}{2})\right)\right].$$

The flux according to formula (4.3) has two peaks, typical profile of flux (4.3) is shown in Fig. 4.2. Figure 4.2 shows that form of flux q_H is highly dependent on heating rates v_T. The first peak can be very narrow at heating rate $v_T = 0.01$ K/min, visible as at heating rate $v_T = 0.05$ K/min, and it can be swallowed up by the second peak as at heating rate $v_T = 1$ K/min. The form flux as Fig. 4.2a can often be found in experimental data, for example [24–26].

The second peak depends on the uniform distribution of hydrogen in the sample central part and it is described by the first mode of solution (4.3), this is shown in paper [14]. The expression for the dependence of the maximum temperature on the heating rate is expressed by formula (4.4), which is a condition on the extremum of the first mode and is like the famous Kissinger formula (4.1).

$$\ln\left(\frac{v_T}{T^2}\right) = \ln\left(\frac{R}{U}D_0\lambda_0^2\right) - \frac{U}{R}\frac{1}{T}. \tag{4.4}$$

We would like to understand connection between the first peak hydrogen flux and high hydrogen concentration near-edge sample. That is, it is necessary to consider diffusion equations (4.2) with initial condition as in Fig. 4.1a with $C_1 = 0$. But we can consider a simpler task at half-infinity (4.5) with initial condition as on Fig. 4.1b, because they have a similar hydrogen flow at the edge

$$\begin{cases} \dfrac{1}{D(t)}\dfrac{\partial C}{\partial t} = \dfrac{\partial^2 C}{\partial x^2}, \\ C(t,x)|_{x=0} = 0, \\ C(0,x) = C_{in}(x), -\infty < x \le 0, \end{cases} \tag{4.5}$$

where the function $C_{in}(x)$ has a form as in Fig. 4.1b.

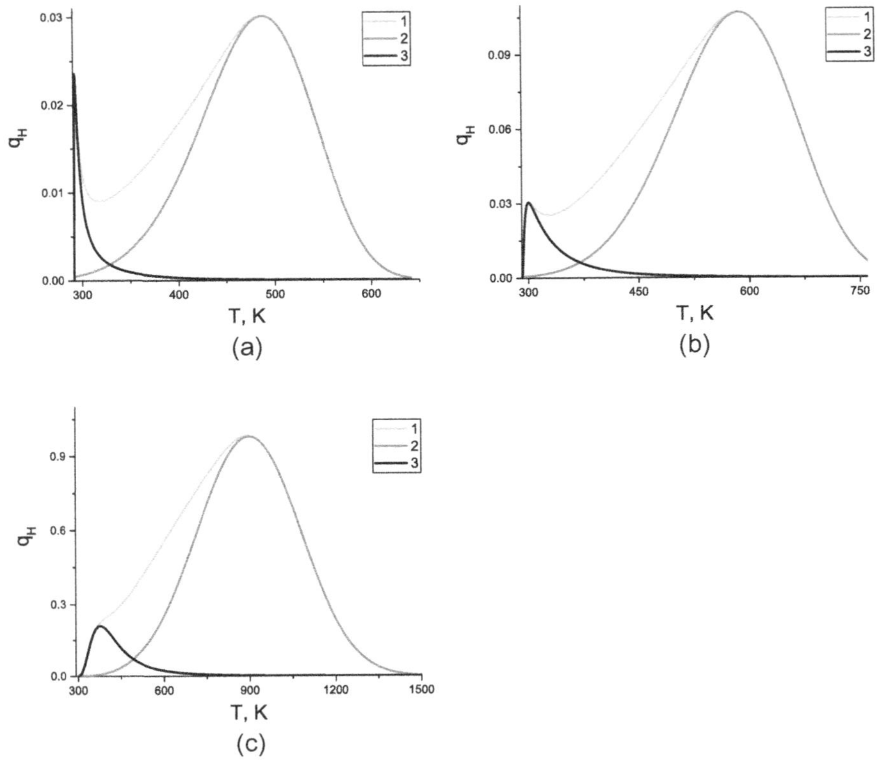

Fig. 4.2 Assessing the effect of the distribution of hydrogen in a sample on the shape of a hydrogen flux. Curve 1 corresponds to formula (4.3), curve 2 corresponds to first term of formula (4.3), and curve 3 corresponds to formula (4.6). **a** $v_T = 0.01$ K/min; **b** $v_T = 0.05$ K/min; **b** $v_T = 1$ K/min. Parameters are shown in Tables 4.1 and 4.2

Table 4.1 Material parameters value

U, kJ/mol	T_0, K	D_0, mm^2/s
30	291	10

Table 4.2 Geometry parameters value

X, mm	X_1, mm	w, mm	C_1/C_2
10	9.8	0.1	1/4

Solution of the system (4.5) can be written using the Fourier transform and hydrogen flux at the edge can be presented as

$$q_H = -D(t)\frac{\partial C}{\partial x}|_{x=0} = -D(t)S(t), \tag{4.6}$$

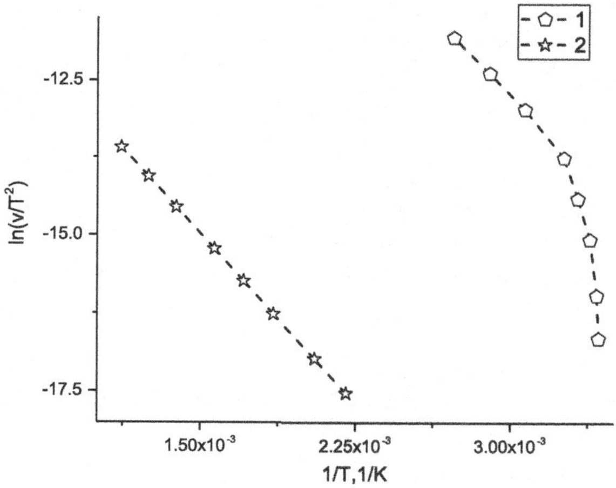

Fig. 4.3 Choo - Lee plot for parameters in Tables 4.1 and 4.2. Line 1 corresponds to Eq. (4.7) for first peak, line 2—(4.4) for second peak

where

$$S(T, v_T) = \sqrt{\frac{\pi}{2}} \int_0^{\infty} qC_{in}^*(q) exp(-q^2 D_0 I(t)) dq,$$

with C_{in}^* is the Fourier sine transformation of the function C_{in}. Comparison of full hydrogen flux and the first term of the original task (4.3) and hydrogen flux of the auxiliary task (4.6) are presented in Fig. 4.2. The flux of original task (4.3) and the flux of the auxiliary task (4.6) coincide at low temperatures. It can be concluded that the first peak corresponds to the desorption near-boundary hydrogen, and expression for the first peak for equations (4.2) coincides with the expression for equations (4.5). The first peak condition of flux (4.3) can be presented as (4.4).

$$\frac{1}{T^2} = \frac{R}{U} \frac{S'(t)}{S(t)}. \tag{4.7}$$

In a simple case, the flux (4.3) of the ordinary diffusion equation (4.2) with a piece linear initial condition has two peaks and dependence extreme temperatures on heating rate is nonlinear in Choo-Lee plot, see Fig. 4.3. The dependence for the second peak, which associated with the exit of bulk hydrogen, is linear in coordinates $1/T$ and $\ln\left(\frac{v_T}{T^2}\right)$. But the dependence for the first peak has two asymptotics, for "low" heating rates is a vertical line, for "higher" one is a line with slope of U/R as for the second peak and nonlinear dependence for middle heating rates.

Typically, samples are tested at several heating rates and data may lay on different sections of the line (4.4). The error of determining the energy can be huge if there is no understanding under what mode of heating these data were obtained: at low, high, or intermediate heating rates. If the experimental points correspond to a high or low heating rate, then the energy value for the first peak will coincide with the energy for the second or it will be very large. The value of energy will be questionable and will only require further research on the data. But if the experimental points belong to intermediate heating rates or to different types of heating rates, then the energy value is arbitrary, but it is reasonable. This will allow making a wrong conclusion about a new type of trap with the corresponding energy.

We can conclude that the assumption that each hydrogen flux peak is associated with activation energy can be incorrect for some cases. Thus it is necessary to consider influence from hydrogen distribution in a sample after pre-charged. To assess the influence of the initial conditions, it is necessary to compare the experimental data and the simulation results of the corresponding equations.

4.3 Simulation and Comparison with Experimental Data

Popular forms of experimental samples are disk or cylinder. Consequently, a solution diffusion equation in axisymmetric cylindrical coordinates is needed for comparable analysis experimental and model data. The diffusion equation in axisymmetric cylindrical coordinates for half sample is shown below:

$$\begin{cases} \dfrac{1}{D(t)}\dfrac{\partial C}{\partial t} = \dfrac{\partial^2 C}{\partial z^2} + \dfrac{1}{r}\dfrac{\partial}{\partial r}\left(r\dfrac{\partial C}{\partial r}\right), \\ C(t,r,z)|_{z=X_z} = C(t,r,z)|_{r=X_r} = 0, \\ \dfrac{\partial C(t,r,z)}{\partial z}|_{z=0} = 0, \\ C(t,r,z)|_{t=0} = C_{in}(r,z), \end{cases} \tag{4.8}$$

where $2X_z$ is a sample height, X_r is a sample radius . $C_{in}(r,z)$ is a piece linear function along the r and z axes. $D(t)$ is a diffusion coefficient, which depends on time $D(t) = D_0 \exp \frac{-U}{RT(t)}$.

System (4.8) is solved numerically, using an semi-implicit finite difference scheme (4.9) with uniform grid in r and z.

$$\begin{cases} \dfrac{r_i}{D(t^{n+1/2})}\dfrac{C_{i,j}^{n+1/2} - C_{i,j}^{n}}{\Delta t/2} = \Delta_r(1/2) + \Delta_z(0), \\ \dfrac{r_i}{D(t^{n+1})}\dfrac{C_{i,j}^{n+1} - C_{i,j}^{n+1/2}}{\Delta t/2} = \Delta_r(1/2) + \Delta_z(1), \end{cases} \tag{4.9}$$

where

$$\begin{cases} \Delta_r(\alpha) = \dfrac{r_{i+1/2}\frac{C^{n+\alpha}_{i+1,j}-C^{n+\alpha}_{i,j}}{\Delta r} + r_{i-1/2}\frac{C^{n+\alpha}_{i-1,j}-C^{n+\alpha}_{i,j}}{\Delta r}}{\Delta r}, \\ \Delta_z(\alpha) = r_i \dfrac{C^{n+\alpha}_{i,j+1} - 2C^{n+\alpha}_{i,j} + C^{n+\alpha}_{i,j-1}}{\Delta z^2}, \end{cases}$$

with superscript means a point in time, for examples, $t^{n+1} = t^n + \Delta t$, $t^{n+1/2} = t^n + \Delta t/2$, subscript corresponds to a coordinate in space $C_{i,j} = C(r_i, z_j)$ is a concentration at the point (r_i, z_j) and $r_{i+1} = r_i + \Delta r$, $r_{i+1/2} = r_i + \Delta r/2$, where $0 \le i < N, 0 \le j < M$.

The calculation is carried out for half of the sample with boundary conditions

$$C_{N,j} = C_{i,M} = C_{i,1} - C_{i,0} = 0, \qquad 0 \le i < N, 0 \le j < M.$$

Our solution strategy (4.9) includes two steps: (i) in the first half of the time step, the flux in the direction of the z-axis is taken at the previous time; (ii) in the first half of the time step, the radial flux is taken at the previous moment in time. The Thomas algorithm is used to solve each equation of system (4.9). Hydrogen flux as the sum of the fluxes from the side surface and two ends is calculated as

$$q_H = 2\pi X_r D^n \sum_{i=0}^{N-1} \frac{C^n_{i+1,M} - C^n_{i,M}}{\Delta r} \Delta z + \\ + 4\pi X_r D^n \sum_{j=0}^{M-1} \frac{C^n_{N,j+1} - C^n_{N,j}}{\Delta z} \Delta r \tag{4.10}$$

Modeling of experimental data can be provided by formulas (4.10). Experimental data are taken from work [23], ferrite-bainite steels circular samples of 20 mm diameter and 1 mm thickness were tested at 3.33, 6.66, 13.3, and 20 heating rate K/min, see Fig. 4.4.

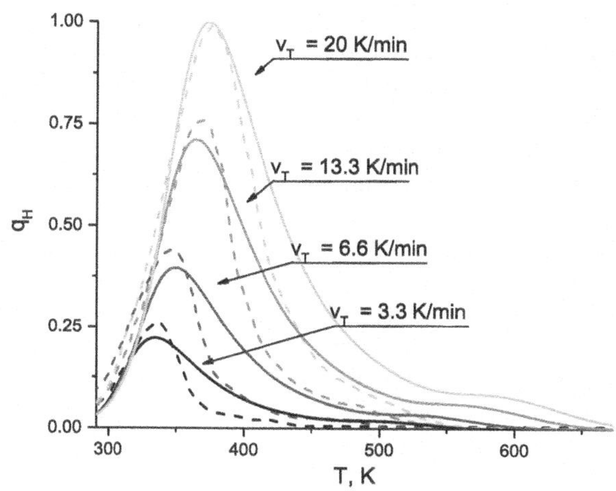

Fig. 4.4 Calculated (by the present model eq. (4.8)) and experimental curves [23] for samples at heating rate 3.33, 6.66, 13.3, 20 K/min. All curves are normalized to the flux maximum value at a temperature of 20 K/min. Parameters value are seen in Tables 4.3, 4.4

Table 4.3 Material parameters value from Ref. [23]

U, kJ/mol	T_0, K	D_0, mm^2/s
42.7	291	230

Table 4.4 Geometry parameters value from Ref. [23]. The subscripts r and z denote the coordinates r and z, respectively

X_r, mm	X_{1r}, mm	w_r, mm	X_z, mm	X_{1z}, mm	w_z, mm	$C_{1r}/C_{r2}/C_{1z}/C_{2z}$
10	9.83	0.1	0.5	0.474	0.024	1/1/1/200

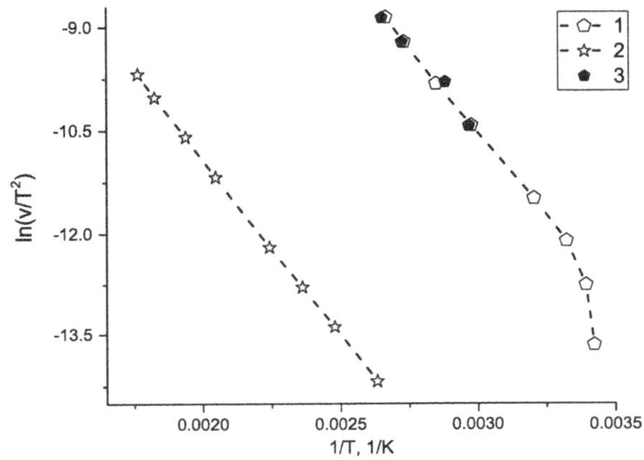

Fig. 4.5 Calculated Choo-Lee plot. Lines 1 and 2 are the numerical data (by the present model Eq. (4.8)), parameters value see in Tables 4.3, 4.4. Line 1 corresponds to equation (4.7) for first peak, line 2—(4.4) for second peak. Black circles (line 3) are experimental data from paper [23]. Additional points are shown for heating rates 0.1, 0.5, 0.5, and 1 K/min

Curves have two peaks, according to our assumption and formulas (4.4, 4.7), the slope in Choo-Lee plot for the second peak determines the activation energy U of equation (4.8). Geometry parameters and D_0 are unknown. They are selected from a fitting of the experimental curve and the curve according to formula (4.10), data for temperature rate 20 K/min is used as a nominal curve. The material and geometry parameters are given in Tables 4.3 and 4.4. A higher concentration of hydrogen at the ends at $z = X_z$ than at surface $r = X_r$ is due to the specifics of sample preparation.

The calculated and experimental profiles for each of rates have similar forms. The slight difference in the quantitative value may be due to the heterogeneity of the samples or to the necessary refinement of the model.

Figure 4.5 shows the dependence of the $\ln(v_T/T^2)$ on $1/T$. The curve corresponding to the second peak is a straight line with a slope coefficient U/R, the curve for the first peak has two asymptotics: vertical for low heating rates and asymptotics with a slope coefficient U/R for high heating rates. According to our resolution in Chap. 4.2, the experimental data [23] corresponds to "high" heating rates. But the dependence for first peaks is more complicated and can be misleading about activation energy for low heating rates than in work [23], examples 0.1, 0.25, 0.5, and 1 K/min . This confirms the importance of understanding the range of heating rates for thermal desorption analysis.

It is necessary to note that each surface gives flux, which consists of two peaks. Real samples are three-dimensional body and they have more than one surface. Consequently, the flux can have more than two peaks, which are related only to the inhomogeneous initial hydrogen distribution in the sample, ignoring the hydrogen trap. In the case under consideration, for thin circular ones, the most important flux is the flux from the surfaces $z = X_z$.

4.4 Conclusion

In this paper, we propose a model describing the effect of the initial condition of hydrogen on TDS. Analytical calculations in the case of one-dimensional Cartesian coordinates show that the edge flux of the ordinary diffusion equation with a piecewise linear initial condition has two peaks, and the Choo-Lee plot for the first peak is a nonlinear function and has two asymptotics. The first peak of the flux corresponds to edge hydrogen, and the second peak corresponds to bulk hydrogen. It highlights the importance of heating rates range for thermal desorption analysis.

Numerical modeling of experimental data shows that the curves obtained at TDS are well described using the diffusion equation and taking into account the non-uniform distribution of hydrogen in the sample. Finite difference scheme is used for solving diffusion equation in antisymmetric cylindric coordinates.

Based on the data obtained, it can be concluded that it is necessary to take into account the effect of the hydrogen distribution in the sample after pre-charged when analyzing the experimental thermal desorption curves.

Acknowledgements This study was supported by the Russian Foundation for Basic Research (project No. 18-08-00201).

References

1. Cailletet, et al.: First report of h embrittlement of metals. Compt. Rend. **58**, 327 (1864)
2. Al-Mufachi, N., Rees, N., Steinberger-Wilkens, R.: Hydrogen selective membranes: a review of palladium-based dense metal membranes. Renew. Sustain. Energy Rev. **47**, 540–551 (2015). https://doi.org/10.1016/j.rser.2015.03.026
3. Akamatsu, T., Kume, Y., Komiya, K., Yukawa, H., Morinaga, M., Yamaguchi, S.: Electrochemical method for measuring hydrogen permeability through metals. J. Alloys Compounds **393**(1–2), 302–306 (2005). https://doi.org/10.1016/j.jallcom.2004.10.007
4. Pisarev, A.A., Ogorodnikova, O.V.: Elementary processes near the interface between hydrogen gas and solid. J. Nuclear Mater. **248**, 52–59 (1997). https://doi.org/10.1016/S0022-3115(97)00201-8
5. Johnson, W.H.: II. On some remarkable changes produced in iron and steel by the action of hydrogen and acids. In: Proceedings of the Royal Society of London **23**(156–163), 168–179 (1875). https://doi.org/10.1016/0025-5416(72)90109-7

6. Louthan, M., Caskey, G., Donovan, J., Rawl, D.: Hydrogen embrittlement of metals. Mater. Sci. Eng. **10**, 357–368 (1972)
7. Bhadeshia, H.K D.H.: Prevention of hydrogen embrittlement in steels. ISIJ Int. **56**(1), 24–36 (2016). https://doi.org/10.2355/isijinternational.ISIJINT-2015-430
8. Kirchheim, R.: Hydrogen solubility and diffusivity in defective and amorphous metals. Progr. Mater. Sci. **32**(4), 261–325 (1988). https://doi.org/10.1016/0079-6425(88)90010-2
9. McNabb, A., Foster, P.: A new analysis of diffusion of hydrogen in iron and ferritic steels. Trans. Metal. Soc. AIME **227**(3), 618 (1963)
10. Oriani, R.A.: The diffusion and trapping of hydrogen in steel. Acta metallurgica **18**(1), 147–157 (1970). https://doi.org/10.1016/0001-6160(70)90078-7
11. Johnson, H.N., Quick, N.A., Kumnick, A.: Hydrogen trapping mechanisms by permeation techniques. Scr. Metall.(United States) **13**(1) (1979). https://doi.org/10.1016/0036-9748(79)90392-2
12. Leblond, J., Dubois, D.: A general mathematical description of hydrogen diffusion in steels-I. Derivation of diffusion equations from Boltzmann-type transport equations. Acta Metallurgica **31**(10), 1459–1469 (1983). https://doi.org/10.1016/0001-6160(83)90142-6
13. Turnbull, A., Carroll, M., Ferriss, D.: Analysis of hydrogen diffusion and trapping in a 13% chromium martensitic stainless steel. Acta Metallurgica **37**(7), 2039–2046 (1989). https://doi.org/10.1016/0001-6160(89)90089-8
14. Belyaev, A.K., Polyanskiy, A.M., Polyanskiy, V.A., Sommitsch, C., Yakovlev, YuA: Multichannel diffusion vs TDS model on example of energy spectra of bound hydrogen in 34CrNiMo6 steel after a typical heat treatment. Int. J. Hydrogen Energy **41**(20), 8627–8634 (2016). https://doi.org/10.1016/j.ijhydene.2016.03.198
15. Hurley, C., Martin, F., Marchetti, L., Chene, J., Blanc, C., Andrieu, E.: Numerical modeling of thermal desorption mass spectroscopy (TDS) for the study of hydrogen diffusion and trapping interactions in metals. Int. J. Hydrogen Energy **40**(8), 3402–3414 (2015). https://doi.org/10.1016/j.ijhydene.2015.01.001
16. Evard, E., Gabis, I., Yartys, V.A.: Kinetics of hydrogen evolution from MgH2: experimental studies, mechanism and modelling. Int. J. Hydrogen Energy **35**(17), 9060–9069 (2010). https://doi.org/10.1016/j.ijhydene.2010.05.092
17. Testi, M., Bartali, R., Crema, L.: Design and optimization of Isochoric Differential Apparatus (IDA) to reduce uncertainty in H2 sorption process measurements. Int. J. Hydrogen Energy **45**(18), 10775–10796 (2020). https://doi.org/10.1016/j.ijhydene.2020.01.190
18. Kissinger, H.E.: Reaction kinetics in differential thermal analysis. Anal. Chem. **29**(11), 1702–1706 (1957). https://doi.org/10.1021/ac60131a045
19. Ma, M., Liang, L., Wang, L., Wang, Y., Cheng, Y., Tang, B., Xiang, W., Xiaohua, T.: Phase transformations of titanium hydride in thermal desorption process with different heating rates. Int. J. Hydrogen Energy **40**(29), 8926–8934 (2015). https://doi.org/10.1016/j.ijhydene.2015.05.083
20. Choo, W., Lee, J.Y.: Thermal analysis of trapped hydrogen in pure iron. Metal. Trans. A **13**(1), 135–140 (1982)
21. Mavrikakis, M., Schwank, J., Gland, J.: Temperature programmed desorption spectra of systems with concentration gradients in the solid lattice. J. Phys. Chem. **100**(27), 11389–11395 (1996). https://doi.org/10.1021/jp9537623
22. Yagodzinskyy, Y., Todoshchenko, O., Papula, S., Hanninen, H.: Hydrogen solubility and diffusion in austenitic stainless steels studied with thermal desorption spectroscopy. Steel Res. Int. **82**(1), 20–25 (2011). https://doi.org/10.1002/srin.201000227
23. Escobar, D.P., Verbeken, K., Duprez, L., Verhaege, M.: Evaluation of hydrogen trapping in high strength steels by thermal desorption spectroscopy. Mater. Sci. Eng. A **551**, 50–58 (2012). https://doi.org/10.1016/j.msea.2012.04.078
24. Takashima, K., Han, R., Yokoyama, K., Funakawa, Y.: Hydrogen Embrittlement induced by hydrogen charging during deformation of ultra-high strength steel sheet consisting of ferrite and nanometer-sized precipitates. ISIJ Int. **59**(12), 2327–2333 (2019). https://doi.org/10.2355/isijinternational.ISIJINT-2019-219

25. Fernandez, J., Cuevas, F., Sanchez, C.: Simultaneous differential scanning calorimetry and thermal desorption spectroscopy measurements for the study of the decomposition of metal hydrides. J. Alloys Compounds **298**(1–2), 244–253 (2000). https://doi.org/10.1016/S0925-8388(99)00620-9
26. Liu, Y., Wang, M., Liu, G.: Hydrogen trapping in high strength martensitic steel after austenitized at different temperatures. Int. J. Hydrogen Energy **38**(33), 14364–14368 (2013). https://doi.org/10.1016/j.ijhydene.2013.08.121

Chapter 5
The Key Role of Dedicated Experimental Methodologies in Revealing the Interaction Between Hydrogen and the Steel Microstructure

Tom Depover and Kim Verbeken

Abstract Understanding the interaction of hydrogen with a steel microstructure is key toward further material development and a potential future hydrogen based economy. A reliable determination of the hydrogen diffusivity and trapping are crucial hereto. Electrochemical permeation is mainly used for the former, while thermal desorption spectroscopy (TDS) is opted for the latter. Combination of both, together with a detailed microstructural evaluation, provides fundamental insights for the engineering of suitable hydrogen traps and diffusion barriers. Finally, in-situ mechanical testing enables to evaluate the hydrogen embrittlement susceptibility and identifies the active deformation mechanism. This chapter summarizes some recent experimental advances and developments of our research group. At first, focus lies on the electrochemical permeation technique, which has been expanded to apply a constant load to the materials to evaluate the influence of stresses on hydrogen diffusivity during in-situ hydrogen permeation. Elastic load on dual-phase steel increases hydrogen diffusion due to crystal lattice expansion, while plastic deformation decreases diffusion due to the formation of lattice discontinuities, e.g., dislocations. Next, the ability to assess hydrogen trapping in austenite-containing materials is critically assessed for a duplex stainless steel, revealing that the TDS data are dominated by hydrogen diffusion in these low diffusivity materials. In-situ interrupted tensile testing on these materials is further complemented with scanning electron microscopy—electron backscatter diffraction analysis. ε- and α'-martensite are found in austenite of the hydrogen-charged tensile specimens, while these martensitic transformation do not show in the uncharged samples. This is, among others, explained by a reduction in stacking fault energy due to the presence of hydrogen. Hydrogen-assisted cracks also initiate in these materials, mainly in the austenite phase. Finally, the hydrogen sensitivity is evaluated for high strength-low ductility Fe–C steels. Due to

T. Depover (✉) · K. Verbeken
Research group Sustainable Materials Science, Department of Materials, Textiles and Chemical Engineering, Ghent University (UGent), Technologiepark 46, 9052 Ghent, Belgium
e-mail: tom.depover@ugent.be

K. Verbeken
e-mail: kim.verbeken@ugent.be

V. A. Polyanskiy and A. K. Belyaev (eds.), *Advances in Hydrogen Embrittlement Study*, Advanced Structured Materials 143, https://doi.org/10.1007/978-3-030-66948-5_5

their brittle nature, a novel in-situ three-point bending setup is developed to evaluate their sensitivity to hydrogen. Hydrogen causes a transition from a microvoid (Fe–0.2C), intergranular (Fe–1.1C), or mixed (Fe–0.4C) fracture surface (air-tested samples), to a hydrogen-induced cleavage fracture appearance. This is accredited by the Hydrogen Enhanced Plasticity Mediated Decohesion mechanism, proposing that hydrogen is preferentially trapped at packet or block boundaries in high carbon steels, while lath martensitic boundaries play a minor role in the crack development.

Keywords Hydrogen embrittlement · Thermal desorption spectroscopy · In-situ mechanical testing · Electrochemical permeation · Hydrogen-assisted cracking

5.1 Introduction

The increased global warming awareness, the limited amount of fossil fuels and concerns about nuclear energy, offer challenges to scientists to provide ecological friendly solutions to deal with these climate-related concerns. As a clean energy vector, a fuel and a feedstock, clean hydrogen has the potential to accelerate the decarbonization of our energy system and of our industrial production pathways. Hydrogen is thus an attractive candidate to replace fossil fuels since its combustion only generates water avoiding the emission of greenhouse gasses. Despite these benefits, hydrogen entails a negative connotation linked to several incidents demonstrating its potential hazard [1], together with the ongoing debate and lack of understanding on the role of hydrogen on the structural mechanical integrity of metals. Besides the development of the hydrogen economy, lowering the fuel consumption of vehicles comprises another approach to make transportation more environmental friendly. Therefore, the use of high-strength steels is encouraged in the automotive industry as it may yield an increased safety together with vehicle's weight reduction, mandatory to meet the rigorous CO_2 emission regulations. Unfortunately, these steels are assumed to be prone to hydrogen-induced degradation [2–4]. Furthermore, steel alloys for the offshore industry are often protected against corrosion by impressed current cathodic protection. Though, when cathodic overprotection takes place, atomic hydrogen can absorb into material's microstructure, causing a degradation of the structural mechanical integrity [5].

Johnson [6] was the first to refer to the detrimental role of hydrogen on the mechanical performance. This pioneering work inspired many researchers, leading to several reference works in the field [7–14]. Recently, the interest in the topic got significantly reinforced by the development of numerous applications involving potential interaction with hydrogen, such as oil and gas pipelines, automotive, storage tanks, offshore structures, and welds [15–17]. The particular detrimental effect of atomic hydrogen on the material's integrity is known as the hydrogen embrittlement phenomenon or hydrogen-assisted cracking. The presence of hydrogen in a material can lead to a reduced ductility, toughness, and ultimate tensile strength, possibly triggering unpredictable failure. The four most cited mechanisms to explain hydrogen embrittlement in non-hydride forming materials are Hydrogen Enhanced DEcohesion (HEDE) [18], Hydrogen Enhanced Localized Plasticity (HELP) [19], Adsorption Induced Dislocation Emission (AIDE) [20, 21], and Hydrogen Enhanced Strain-Induced Vacancy (HESIV) [22]. HEDE considers a decrease in cohesive bond strength between the metal atoms in the presence of hydrogen, leading to brittle crack propagation under tensile load. HELP proposes an increased dislocation mobility due to the presence of hydrogen, causing highly localized plastic deformation and finally accelerated failure [23–25]. AIDE combines HEDE and HELP, as it comprises localized plasticity, though the theory proposes that this localized plasticity takes place close to the surface at regions of stress concentrations, e.g., crack tips, including nano-voids ahead. Finally, HESIV proposes that the focus regarding the crucial factor for degradation is shifted from hydrogen to vacancies. The theory states that hydrogen promotes the formation of vacancies upon straining and thus reduces the resistance against ductile crack growth. Nevertheless, although the detrimental hydrogen effect was already discussed in [6] and several mechanisms have been proposed so far, no full comprehension of the phenomenon governing the observed ductility loss has been achieved and complex microstructures complicate our understanding. Conclusively, the potential and safe application of hydrogen still offers major challenges to the materials engineer within the frame of a clearly renewed interest in this research field.

A dedicated experimental methodology is required to gain fundamental understanding on the role of hydrogen on the structure/property relationship. Generally, the presence of hydrogen is known to deteriorate the mechanical performance of steel. However, a critical hydrogen concentration is required to initiate and propagate a crack in the metal microstructure. The susceptibility to hydrogen-induced degradation thus depends on the amount of hydrogen able to diffuse to specific, e.g., stressed, regions in the microstructure. Though, besides the hydrogen concentration, the ability to reach these critical zones is a key parameter as well. This ability is described by the hydrogen diffusion coefficient. Depover et al. [26] visualized the role of hydrogen diffusion by inspecting the fracture surfaces of in-situ charged tensile specimens, tested at different crosshead deformation speeds. The calculated hydrogen diffusion distance, i.e., penetration depth from the edges, is depending on the applied deformation speed. These distances matched perfectly well with the detected transition between hydrogen-induced brittle and ductile fracture features on the fracture surface. The prominent effect of hydrogen diffusivity was further established by in-situ hydrogen tensile testing. Generally, the ductility loss was more pronounced at lower deformation speeds [4, 27], since hydrogen had more time to diffuse and accumulate at critical regions. Hence, the critical hydrogen concentration for crack initiation was reached faster. Hydrogen can indeed accumulate at hydrostatic stress fields ahead of, e.g., a notch or crack tip. Depover et al. [28] evaluated the influence of the local hydrostatic stress state and the resulting local hydrogen concentration on the hydrogen-induced degradation. The maximal concentration of hydrostatic stress and related hydrogen concentration was located in front of the notch tip. When inspecting the fracture surface of the dual-phase steel samples, the present alumina inclusions in this material showed significant embrittlement in the region where hydrostatic stresses and hydrogen concentration peaked. Moreover, hydrogen-induced fish-eyes were found surrounding these inclusions, only at these specific regions of increased hydrostatic stress. Lastly, the synergetic effect of both hydrogen content and hydrogen diffusivity was confirmed in [27]. This work nicely demonstrated that the link between the amount of hydrogen and the hydrogen embrittlement degree is incomplete without integrating hydrogen diffusion into the discussion.

Besides the importance of hydrogen diffusion, the concept of hydrogen trapping offers significant opportunities to tweak the material design toward an improved resistance against hydrogen embrittlement. An important tool to evaluate the hydrogen trapping characteristics is thermal desorption spectroscopy (TDS). This technique is used to evaluate the distribution of hydrogen within a crystal lattice and its discontinuities [29–32]. Hence, hydrogen trapping at microstructural features and defects can be considered. Every type of trap is characterized by a particular desorption activation energy (E_a), being the energy needed for hydrogen to be released from the trapping site. Desorption activation energies can be determined by the Kissinger equation [33] by TDS measurements at different applied heating rates, as introduced by Lee and co-workers [34–36]. However, some recent literature debated the restrictions of the Kissinger equation, especially during diffusion controlled processes, as the possible delay by slow hydrogen diffusion is neglected in the referred approach [31, 37]. Lattice diffusion in body centered cubic (bcc) iron shows a low E_a (8 kJ/mol [38]), while microstructural discontinuities entail significantly higher binding energies (20–60 kJ/mol [29, 31, 32, 39–42]), while precipitates can even reach higher binding energies [30, 43–48]. Consequently, the E_a can be determined in a trustworthy way by means of the Kissinger equation, based on TDS desorption profiles at different heating rates. However, lattice diffusion in face centered cubic (fcc) iron entails a much higher E_a (51–55 kJ/mol). Since the binding energy between hydrogen and microstructural defects is similar to bcc iron, this largely impacts the capacity of TDS to characterize microstructural trapping in fcc metals [49].

Finally, in-situ mechanical testing allows the evaluation of the hydrogen embrittlement susceptibility. By interrupted in-site tensile testing, the active deformation mechanism can be identified by scanning electron microscopy—electron backscatter diffraction (SEM-EBSD) analysis. Laureys et al. [50–52] discussed this specific topic in TRIP steel where retained austenite transformed to α'-martensite upon tensile straining, creating crack initiation sites inside the martensite islands. The martensitic phase is considered to be highly susceptible to hydrogen embrittlement and thus the hydrogen-enriched martensite is very prone to hydrogen-induced cracking. Depover et al. [53] recently verified these observations during in-situ SEM micromechanical testing. Although conventional in-situ hydrogen tensile testing allows a quick and reliable screening method to evaluate the sensitivity to hydrogen embrittlement, the method is not suitable for all materials. This was demonstrated in [27], where lab cast Fe-0.2C and Fe-0.4C alloys were prepared to compare the behavior of different microstructural constituents, i.e., ferrite, pearlite, bainite, and martensite, induced by dedicated heat treatments. Conventional in-situ testing yielded trustworthy results for most of these prepared alloys. Only for the Fe-0.4C materials with a martensitic microstructure, the hydrogen-charged tensile specimens broke upon clamping in the tensile machine. For these high strength-low ductility materials, conventional tensile testing is no longer appropriate to evaluate the proneness to hydrogen. An alternative methodology is, therefore, needed.

This work provides an overview of some of our recent work on the interaction of hydrogen with a steel microstructure. At first, the effect of constant tensile load on the hydrogen diffusivity will be characterized by electrochemical hydrogen permeation experiments during in-situ constant loading on dual-phase steel. Then, a critical assessment of the evaluation of hydrogen trapping of low hydrogen diffusion materials will be presented. For this purpose, duplex stainless steels will be characterized by TDS, combining both ferrite (bcc) and austenite (fcc) to assess the difficulties in interpreting the TDS data with respect to hydrogen trapping. Furthermore, the role of hydrogen on the active deformation mechanism in these materials will be considered as well by interrupted in-situ tensile testing combined with SEM-EBSD characterization. The initiation for hydrogen-assisted cracking will be assessed as well. Finally, the development of a novel in-situ bending technique to evaluate the hydrogen embrittlement sensitivity of brittle materials will be introduced based on lab cast Fe–C alloys, while the operational hydrogen embrittlement mechanism in these materials will be addressed based on SEM fractography.

5.2 Experimental Procedure

5.2.1 Materials Characterization

5.2.1.1 Commercial Steel Grades

The commercial steel grades considered in this work are dual-phase (DP) steel and duplex stainless steel 2205. The first considered material is the DP steel, consisting of ferrite matrix with 23.6% of martensite. The grain sizes of the ferritic and martensitic grains are about 7 μm and 2 μm, respectively. More details can be found in [26]. The thickness of the DP steel sheets was 1.1 mm, reached after hot and cold rolling, followed by subsequent annealing via industrial annealing parameters required to obtain the desired microstructure. The chemical compositions is presented in Table 5.1. The DP ferrite-martensite microstructure results in a good combination of strength and ductility, making them attractive steel grades for automotive applications. However, the mechanical contrast between both phases contains specific hazards, which are even more pronounced in the presence of hydrogen [54–56]. Therefore, this grade deserves further attention toward its hydrogen characteristics under an applied mechanical load. Inclusions were also detected in this material, identified as alumina particles by energy dispersive X-ray analysis. Unfortunately, these oxides were not homogeneously dispersed. Moreover, tensile stresses can concentrate at incoherent regions, i.e., a sharp edge or tip of the inclusion, creating a triaxial stress field where hydrogen will preferentially accumulate [57]. Therefore, a small scatter on the permeation data can be expected.

Table 5.1 Chemical compositions commercial steel grades (in wt%)

Wt%	C	Cr	fNi	Mo	fMn	Si	Other
DP steel	0.08	0.568	0.025	0.005	0.948	0.129	Cu 0.019, P 0.037, V 0.005, S 0.003, Al 0.028
UNS S32205	0.022	22.85	5.50	3.07	1.81	0.32	Cu 0.200, P 0.027, Co 0.162, N 0.173

An as-received plate of UNS S32205 duplex stainless steel with an initial thickness of 0.8 mm was studied as well. The chemical composition is included in Table 5.1. An austenite fraction of 53 $\pm$ 2% was measured by magnetic measurements (Feritscope FMP30). The grains are elongated along the rolling direction with an average thickness of 1.49$\pm$0.1 μm. More microstructural details can be found elsewhere [58, 59]. Duplex stainless steels are two-phase materials consisting of γ-austenite (fcc) and α-ferrite (bcc), showing a combination of excellent mechanical properties and corrosion resistance. The two-phase microstructure causes a complex plastic deformation behavior. Ferrite typically deforms by deformation slip due to its high amount of possible slip systems, whereas austenite can deform by different mechanisms, i.e., dislocation slip, twin formation and martensitic transformations depending, among others, on its stacking fault energy (SFE). The role of hydrogen on the active deformation mechanism in these duplex stainless steel will be evaluated by interrupted in-situ tensile testing.

5.2.1.2 Lab Cast Fe–C Alloys

Three Fe–C alloys with a different carbon content were lab cast and synthesized. The chemical composition of these generic alloys can be found in Table 5.2. The Fe–C alloys were processed by vacuum-induction melting under an argon atmosphere and hot rolled at 1100 °C till final thickness of 1.6 mm. The materials were then austenitized (50 °C > A_{c3}) for 20 min, followed by a brine water (7% NaCl) quench to obtain a fully martensitic structure. The alloys were used in their as-quenched condition to avoid the formation of cementite. More details can be found elsewhere [60]. Each alloy consisted of a martensitic structure (cf. Fig. 5.1), although Fe–1.1C had about 10 vol.% retained austenite, as revealed by X-ray diffraction [60]. Fe–1.1C also contained quench cracks, mainly propagating along the prior austenitic grain boundaries (cf. Fig. 5.1 (Fe–1.1C)). These quench cracks appear due to thermal stresses in steels having a carbon content >0.7 wt.%, since the appearance of plate martensite is necessary [61]. Moreover, when the carbon content exceeds the eutectoid composition, cementite may form along the austenitic grain boundaries during the hot rolling stage, which also facilitates cracking.

Table 5.2 Chemical compositions generic alloys (in wt%)

Wt%	C	Other (N, S, Al)
Fe–0.2C	0.222	<10 ppm
Fe–0.4C	0.421	<10 ppm
Fe–1.1C	1.111	<10 ppm

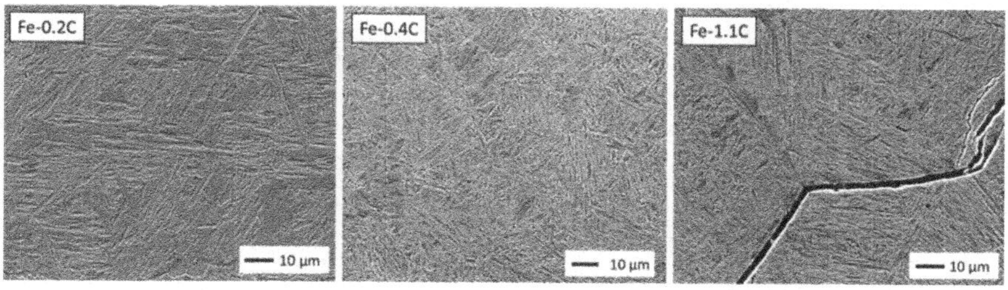

Fig. 5.1 SEM images of cross-sections of the Fe–C samples [60]. Fe–1.1C showed quench cracks. Copyright 2020, Elsevier

5.2.2 Determination of Total Hydrogen Content by Melt Extraction

Melt extraction is used to determine the hydrogen saturation level. Hydrogen was introduced in the alloys by electrochemical charging, using a 1 g/L thiourea 0.5 M H_2SO_4 solution at a current density of 0.8 mA/cm^2, for different times depending on the material of interest. The total hydrogen concentration was determined by melt extraction with an impulse furnace for heating to 1550 °C and a thermal conductivity detector for hydrogen detection. Multiple measurements were done on the duplex stainless steel and the Fe–C alloys for each applied charging time to obtain reproducible results.

5.2.3 Determination of Hydrogen Trapping Capacity Bt Thermal Desorption Spectropscopy

The hydrogen trapping sites and their activation detrapping energy were determined by performing TDS analysis for the duplex stainless steel grade. An infrared furnace to gradually heat the sample up to 950 °C and a quadrupole mass spectrometer was used for simultaneous hydrogen detection. Disc-shaped specimens were tested with a diameter of 20 mm. The thickness of this material was reduced to 0.3 mm with the same surface finish as for the melt tests. The samples were hydrogen charged, similarly as described above and two different charging times were applied, i.e.,

1 day and 15 days. Five different heating rates were applied: 900, 600, 50, 35, and 20 K/h. The method based on the work of Lee and co-workers [34–36] was used to determine the E_a of hydrogen traps related to the peaks observed in the TDS spectra. Equation (5.1) is a simplification of the original formula of Kissinger [33]:

$$\frac{\mathrm{d}\ln(\Phi/T_{\max}^2)}{d(1/T_{\max})} = \frac{E_a}{R}, \tag{5.1}$$

where Φ is the heating rate (K/min), $T_{\max}$ (K) the TDS peak temperature, E_a (J/mol) the detrapping activation energy for the specific H trap associated with $T_{\max}$ and R ($JK^{-1}mol^{-1}$) the universal gas constant. After TDS measurements using different heating rates, deconvolution of the results and determining the corresponding peak temperatures for a trap, plotting $\ln(\Phi/T_{\max}^2)$) versus $(1/T_{\max})$ allows to obtain the E_a corresponding to that specific trap.

The experimental TDS spectra of this duplex steel was complemented with a numeric diffusion model, for which both a homogeneous and heterogeneous microstructural model was developed to simulate the experimental desorption spectra [62]. Hydrogen diffusion according to Fick's law was implemented for this purpose; however, trapping was not included to verify the contribution of the latter to the experimental TDS profile. The first model assumed that the material was homogeneous with one over-all, temperature-dependant, hydrogen diffusion coefficient and an average hydrogen solubility. The second model considered the heterogeneous microstructure of the duplex stainless steel and simulated the material as a combination of two phases with their own hydrogen diffusion coefficient and solubility. On the interface of both phases, local equilibrium was assumed with consideration of a partitioning constant. In this work, the first model will be used. By comparing this model with the experimental TDS data, insight in the contribution of hydrogen diffusion to TDS curves can be obtained. Hence, the potential impact of hydrogen trapping on experimental TDS curves can be reflected.

Homogeneous hydrogen diffusion was simulated using Fick's 2nd law in its one-dimensional form [31, 63]:

$$\frac{\partial C}{\partial t}(x,t) = D(T(t))\frac{\partial^2 C}{\partial x^2}(x,t). \tag{5.2}$$

With $C(x,t)$, the position (x) and time (t) dependent hydrogen concentration [wppm], D the temperature dependent hydrogen diffusion coefficient [m^2/s], T the temperature [K], t the time [s] and x the position [m]. The diffusion coefficient is thus dependent on the time since temperature and time are linked by the constant applied heating rate. Both charging and discharging processes were simulated to reproduce the experimental TDS profiles. For charging, the initial concentration profile was zero throughout the sample's thickness. The temperature was kept constant. The boundary condition for the charging simulation was

$$C(0,t) = C(d,t) = C_s, \quad t > 0. \tag{5.3}$$

With d the thickness [m] and C_s the equilibrium surface concentration for the given parameters [wppm]. Since the potential did not differ during hydrogen charging, constant concentration boundary conditions were selected. The differential equations were numerically evaluated by Forward-Time-Central-Space finite differentiation (spatial discretion was 6 μm). The time discretion was selected based on the Von Neumann stability criterion, i.e., $0.2 \cdot \frac{\frac{1}{2}dx^2}{\max(D(t))}$ with a maximal value of one second, to assure a precise and stable simulation. For discharging, the concentration profile after charging was used as initial state. The boundary condition for discharging was

$$C(0, t) = Ct(d, t) = 0, \quad t > 0. \tag{5.4}$$

Identical spatial and time discretion were selected as for the charging. To attain a signal that eventually can be compared with the experimental TDS spectra, the flux J [wppm/s] was assessed as

$$J(t) = -\frac{d\overline{C}(t)}{dt} \tag{5.5}$$

with $\overline{C}$ the time-dependent average concentration [wppm]. As discharging involves increasing temperatures, the relationship between the diffusion coefficient and temperature should be incorporated. Turnbull et al. [64] reported $D_{\text{eff}}(t) = 2.8 \cdot 10^{-8} \exp\left(-\frac{39300}{RT(t)}\right)$ m^2/s for 2205 duplex stainless steel. The temperature is linearly related to the time via the applied heating rate. This equation was hence used in the initial simulations.

5.2.4 *Determination of Hydrogen Diffusion Coefficient for Hydrogen Permeation Under Constant Load*

The electrochemical permeation technique was used to determine the hydrogen diffusion coefficient, based on the Devanathan and Stachurski method [65]. A permeation setup was combined with an external loading device (cf. Fig. 5.2). It consisted of two double glass cells with the sample clamped in between. In both cells, a three-electrode system was present, with a working electrode (WE), Pt counter electrode (CE), and Hg/Hg_2SO_4 reference electrode (RE, +650 mV vs. the standard hydrogen electrode, SHE). The three electrodes were connected to a potentiostat in both the cathodic and anodic cell. The temperature of the electrolyte, 0.1 M NaOH in both cells, was kept constant at 25 °C by circulating water in the outer layer of the cells. N_2 gas was blown into the air-tight compartments to reduce the oxygen content and hence minimize unwanted oxidation. The sample's thickness was about 1 mm, as such, hydrogen diffusion was bulk controlled. The methodology is elaborated in more detail in [66].

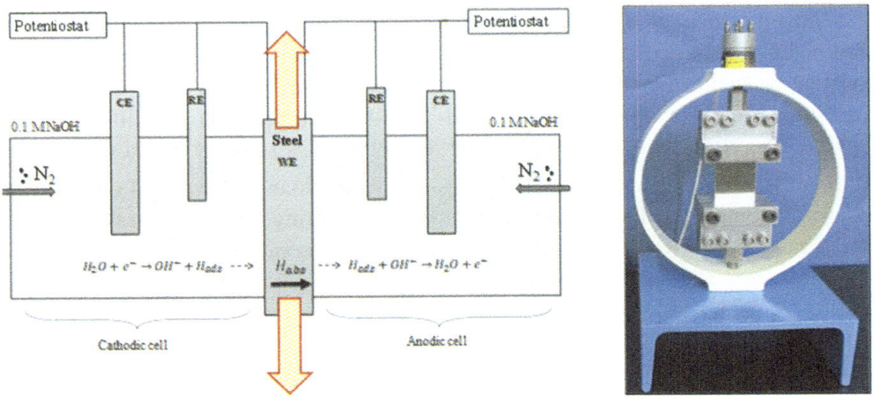

Fig. 5.2 Schematic representation of the electrochemical permeation setup (left) adjusted to allow mechanical loading by a proof ring (right). CE = counter electrode, RE = reference electrode, WE = working electrode (steel) [72] Copyright 2020, Elsevier

It was opted to not apply a Pd layer at the exit side of the sample since during plating process, defects and/or oxides at a metal/Pd interface can be easily introduced. These heterogeneities affect the hydrogen diffusion in an uncontrolled manner, which is not desirable. Moreover, reliable permeation results can be obtained without the use of Pd as well [67–71]. Instead, a passive iron oxide layer was imposed by applying a potential of −500 mV versus RE before starting up the actual experiment. Before this passivation step, the oxygen content in the solution was diminished by blowing N_2 gas through the electrolyte. The anodic cell was filled with the solution by a peristaltic pump and constant circulation was maintained throughout the experiment. Once a stable background current density (<100 nA/cm^2) was achieved, the hydrogen permeation test started, for which the cathodic cell was filled with deaerated NaOH solution under galvanostatic control (−3 mA/cm^2). The hydrogen oxidation current density versus time was then further processed to determine the hydrogen diffusion coefficient, as described below.

To apply a constant load during the in-situ permeation test, the sample was first mounted in the proof ring. Then, the permeation setup was fitted around the specimen. An average thickness of five different measurements in the zone of interest was used to calculate the diffusion coefficient. Hydrogen permeation experiments were performed with a variety of loads on the DP steel. The stress level is expressed as a fraction of the yield strength (YS) and is mentioned in the sample code. Stresses of 60, 80, 100, 120, and 140% of the YS were applied.

The apparent diffusion coefficient ($D_{app,fit}$) was determined by fitting the theoretical permeation transient, based on Fick's law, with the first half of the normalized permeation transient. This method was selected as the experimental transient started to deviate from the theoretical one at about a normalized current of 0.5. This deviation was clarified by the deposition of metal cations present in the electrolyte, as elaborated elsewhere [66]. Using Fourier series and the right boundary conditions, Fick's 1st and 2nd law could be solved. In this work, the experimental curves were fitted with the galvanostatic model (constant flux, CF) which is expressed by Eq. 5.6 [73, 74]. According to the ASTM international standard, a value of at least n = 6 is recommended [75]. In this work, a value of n = 10 was used.

CF model:

$$\frac{J}{J_\infty} = 1 - \frac{4}{\pi}\sum_{n=0}^{10}\frac{(-1)^n}{2n+1}\exp\left(-\frac{(2n+1)^2\pi^2 Dt}{4L^2}\right). \tag{5.6}$$

The first step in the fitting procedure was to plot a J_{norm} versus time diagram. This was then fitted to the CF theoretical model. The optimal fit was obtained by minimizing the total error (Eq. 5.7) between the experimental and theoretical curve. Therefore, D was varied using the solver Add-in in Excel.

$$\text{Total error} = \sum\left(\frac{J_{exp} - J_{back}}{J_{exp,\infty} - J_{back}} - \frac{J_{th} - J_{back}}{J_{th,\infty} - J_{back}}\right)^2 \tag{5.7}$$

5.2.5 Determination of the Degree of Hydrogen-Induced Mechanical Degradation

By comparing mechanical tests performed in air with in-situ tests done on electrochemically hydrogen-charged specimens, the impact of hydrogen on the mechanical performance can be assessed. The samples were electrochemically hydrogen charged by the same method as described above.

At first, the impact of the hydrogen presence on the mechanical behavior of duplex stainless steel was tested by conventional tensile testing. The length direction of the tensile samples was chosen parallel to the rolling direction. Tensile tests were performed at a constant crosshead displacement speed of 0.6 mm/min (strain rate of $1 \cdot 10^{-3} s^{-1}$). Uncharged tests (in air) were done as a reference. Hydrogen-charged tensile specimens were tested in-situ, i.e., with continuous hydrogen charging during the tensile test, after pre-charging for 24 or 168 h. Theoretical calculations of the hydrogen concentration profiles through the thickness of the sample are shown in [58]. These calculations assumed a homogeneous material with one average hydrogen diffusion coefficient and revealed that both charging times did not result in hydrogen saturation. Multiple interrupted tensile tests were done to study the role of hydrogen on the active deformation mechanism of this material, for which both in-situ and ex-situ charging was considered. Furthermore, additional interrupted tensile tests were done, at higher strain levels, to evaluate the initiation and propagation of hydrogen-assisted cracks (HAC). SEM-EBSD was used to study the deformation mechanism and HAC characteristics. The used SEM (FEI QUANTA FEG 450) operated at an

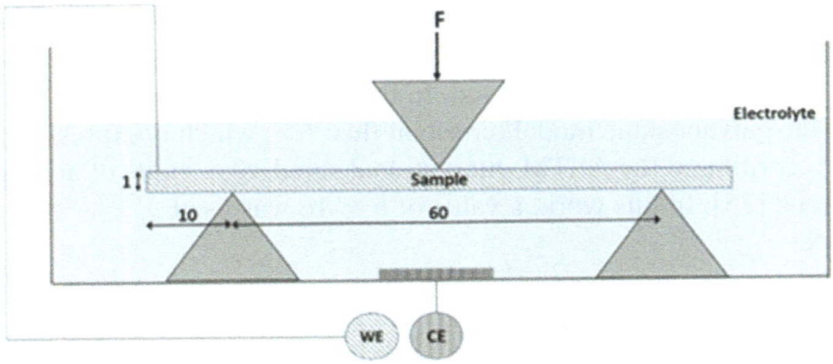

Fig. 5.3 Schematic representation of the in-situ bending technique [60] Copyright 2020, Elsevier

accelerating voltage of 20 kV and a spot size of 5 nm. The specimen was tilted to 70 °C for EBSD measurements. The step sizes used varied between 0.1 and 0.02 µm on a hexagonal grid depending on the used magnification. TSL-OIM Data analysis V7.3 software was used for post-processing and analysis of the crystallographic orientation data.

Secondly, martensitic Fe–C alloys containing a carbon content >0.4 wt% and charged with hydrogen, failed already in the elastic regime during conventional tensile testing [27], as introduced in the introduction part. For these materials with limited ductility, a novel in-situ bending setup was developed to allow the evaluation toward their hydrogen sensitivity. A schematic representation of the setup is presented in Fig. 5.3. During a bending test, a constant transverse loading speed is applied on a beam-shaped sample. The maximum bending stress (σ_{nom}) in a straight beam of height a, thickness B, an applied force F and a moment arm of length z can be calculated by the simple beam theory as

$$\sigma_{nom} = \frac{6Fz}{Ba^2}. \tag{5.8}$$

However, the validity of this formula is restricted to the elastic regime. In the plastic region, some deviations may occur, but since this work deals with materials with limited ductility, the formula was used for generating both the elastic and plastic region. The bending tests were done for three testing conditions: without hydrogen charging (air), with hydrogen charging during the bending test (in-situ), and pre-charged with hydrogen until saturation and continued in-situ charging during the bending test (saturated in-situ). Two different crosshead deformation speeds (1 mm/min and 0.1 mm/min) were applied. The samples were ground and polished till 1 mm to obtain a defect-free surface and to avoid surface-induced stress concentrations during testing. The samples' thickness was identical for both the mechanical and the hydrogen testing methods, assuring a reliable comparison between the different testing methods.

5.3 The Interaction of Hydrogen with Advanced High-Strength Steels

5.3.1 The Effect of Constant Tensile Load on the Hydrogen Diffusivity of DP Steel by Permeation

A constant tensile load of 60, 80, or 100% of the YS of the DP steel was applied prior to the start of the permeation test. This load sustained during the permeation experiment, corresponding to stress values of 195 MPa, 260 MPa, and 325 MPa, respectively. $D_{app,fit}$, presented in Fig. 5.4a, is obtained by fitting the first half of the normalized current, as elaborated above. The first half of the normalized permeation transients are plotted in Fig. 5.4b. The small scatter can be clarified by the presence of the alumina inclusions, cf. materials characterization section. When comparing the 60%YS and the 80%YS condition, the increased applied load caused an increased diffusivity. Zhao et al. [73] detected a similar increase of D_{app} with imposed elastic stresses. On the one hand, elastic tensile stresses increased the volume of the unit cell and its interstitial sites, which is favorable for hydrogen diffusion [73]. On the other hand, the applied stress increased the hydrogen concentration of the subsurface (C_0). This could also be responsible for the higher $D_{app,fit}$, as an increased $D_{app,fit}$ with increasing C_0 was detected in [76–78]. The cathodic reduction reaction rate at the input surface is higher at a stressed surface leading to a higher hydrogen concentration in the subsurface of the sample.

Though, at 100% of the YS, the diffusivity decreased again reaching similar values as the 0%YS condition (Fig. 5.4a). Since the YS was determined by the 0.2% strain offset method, the stress–strain curve already clearly deviated from the linear elastic part before this determined YS was reached [72]. Moreover, hydrostatic tensile stresses may concentrate surrounding the present inclusions, leading to a local triaxial stress fields where hydrogen can accumulate [28, 57]. As such, some micro-plasticity was locally induced and, correspondingly, dislocations nucleated showing a retardant effect on the diffusivity. The increase in dislocation density with applied (plastic)

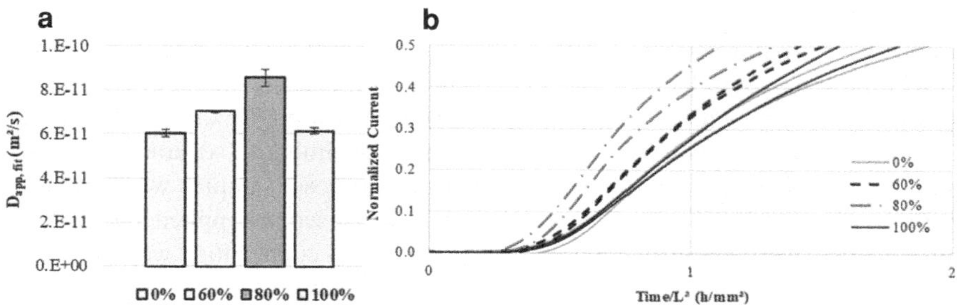

Fig. 5.4 Apparent diffusion coefficients for DP steel under constant elastic stress (% YS), with 0% as reference (**a**). First half of the normalized permeation curves for the different loading conditions (**b**) [72] Copyright 2020, Elsevier

deformation was confirmed in our previous work on this steel grade [39]. Here, it was established that the ferrite grains took most of this plastic deformation. Furthermore, several authors confirmed that dislocations can be considered as hydrogen traps reducing the hydrogen diffusion [79, 80]. This additional trapping compensated for the lattice expansion resulting from the elastic stresses, which are indeed still present as the sample is sustained under constant load. These findings were in agreement with the previously published results, where a decrease in diffusion was observed when applying a stress close or equal to the YS. Again, the generation of microstructural defects by plastic deformation compensated for the lattice expansion [57, 73].

Furthermore, the formation of vacancies and vacancy clusters might also play a role as an increase in vacancy concentration by the annihilation of dislocations during plastic deformation was verified in [81, 82]. Moreover, the formation of vacancies and vacancy clusters is enhanced in the presence of hydrogen [83]. Hydrogen was confirmed by positron annihilation spectroscopy to enhance the generation of vacancies during plastic deformation [84], which was in agreement with TDS data demonstrating an increase in hydrogen absorption by straining [85]. Molecular dynamics and cluster dynamics simulations showed that a hydrogen-vacancy complex is not absorbed by dislocations sweeping through the lattice, unlike a lattice vacancy [86]. When metals undergo plastic deformation in the presence of hydrogen, the simulations demonstrate that the dislocation motion and dislocation–dislocation intersections produce a large amount of hydrogen-vacancy complexes. Under these concentrations, the complexes prefer to grow by absorbing additional vacancies and act as nuclei for nano-voids formation. Vacancies and vacancy clusters are stable and grow easier due to their lower formation energy with the help of trapped hydrogen [87]. These vacancies and vacancy clusters can hinder diffusion by trapping hydrogen as well [88] Therefore, they presumably also contributed to the decreased diffusivity of the 100% YS condition.

The $D_{app,fit}$ when a constant load in the plastic regime was applied (including the elastic part), are plotted in Fig. 5.5a, including the reference sample, DP0%. The decreased diffusivity, detected for the imposed load of 120 and 140% of the YS, was caused by additional lattice defects induced by increased plastic straining. Consequently, the impact of the plastic stress on the hydrogen diffusion considerably exceeded the effect of the expanded lattice, as both elastic and plastic stresses are present under these loading conditions. The decrease in diffusion due to plastic stresses was also discussed in [89–91]. When a constant load was applied in the plastic regime, self-evidently, the entire contribution of the elastic strain also remained present. As mentioned before, elastic stresses caused a higher diffusivity. Hence, a supplementary proof-of-concept test was designed aiming to separate the contribution of the elastic and plastic stresses. For this purpose, samples were tested by applying a load of 120% of the YS, which was then released (no applied load) prior to the start of the permeation test. As such, the elastic stress contribution was eliminated from the permeation result and only the permanent plastic contribution endured. The related average $D_{app,fit}$ is indicated in Fig. 5.5b as "120%*". A further reduction in

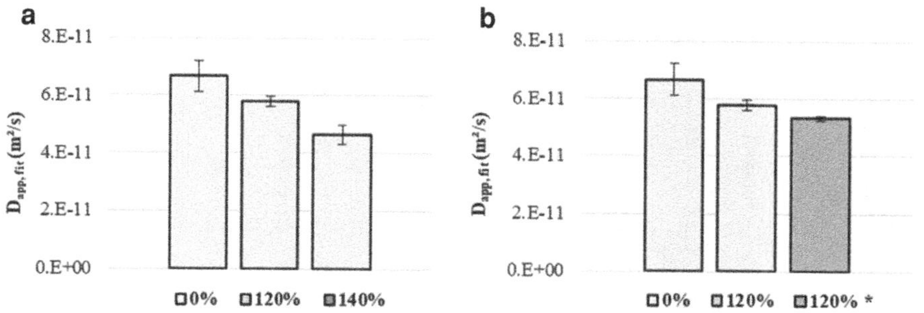

Fig. 5.5 Apparent diffusion coefficients for DP steel under constant applied stress in the plastic regime 120% and 140% YS, with 0% as reference condition (**a**). 120%* was unloaded prior the permeation experiment to confirm the effect of elastic stress (**b**) [72] Copyright 2020, Elsevier

diffusion was observed when the elastic stresses were eliminated. This demonstrated once more that elastic stresses caused a higher diffusion, whereas the microstructural changes by plastic deformation accounted for a decrease in diffusivity.

5.3.2 Revealing the Role of Hydrogen in Duplex Stainless Steels

5.3.2.1 Critical Assessment of Hydrogen Trapping of Duplex Stainless Steel Evaluated by TDS

Figure 5.6 present the hydrogen uptake capacity versus applied charging time by melt extraction tests, including the fit of the analytical solution of Fick's 2nd law. The saturation level of this material was about 702 wppm. The effective diffusion coefficient was determined to be $9.94 \cdot 10{-}15$ m^2/s, which is in good agreement with literature [92, 93]. Figure 5.7 shows the obtained TDS data; both fast heating rates (600 and 900 K/h, cf. Fig. 5.7a) and slow heating rates (20, 35 and 50 K/h, cf. Fig. 5.7b) were applied. The latter because of a potential enhanced peak separation between diffusion and trapping processes. The spectra of the samples charged for 15 days showed a different shape, dependent on the applied heating rate. The fast heating rates (15d H) resulted in one asymmetric peak with a rather flat top. The slow heating rates (15d H) caused a small peak followed by a two-step peak. The samples charged for 1 day (1d H) revealed one main peak followed by a shoulder finishing at about the same temperature as the corresponding 15 days charged sample. This was observed for both the fast and slow heating rates. However, the 1 day charged specimen tested at 50 K/h showed an additional small peak in the beginning as was also the case for the 15 days charged sample at this heating rate (cf. Fig. 5.7b).

The desorption activation energy was determined based on the experimental data, as described above, and resulted in an average E_a of 43.4 kJ/mol. This experimen-

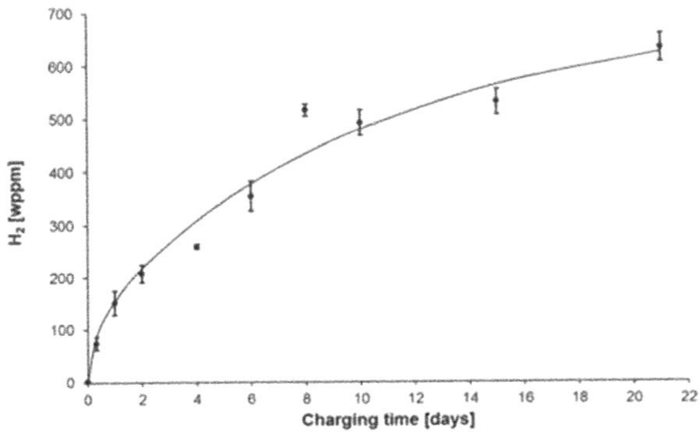

Fig. 5.6 Hydrogen content versus charging time with fit based on analytical solution of Fick's 2nd law [62] Copyright 2020, Elsevier

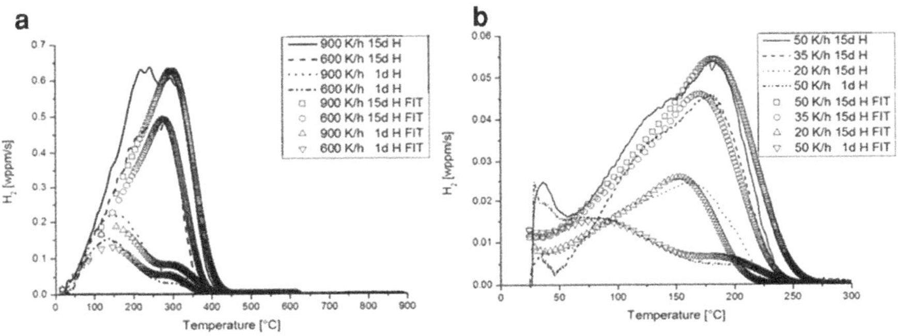

Fig. 5.7 TDS data of 2205 duplex stainless steel for different charging times and heating rates (fast in (**a**), slow in (**b**)). Fit of the homogeneous diffusion model to the experimental TDS data included [62] Copyright 2020, Elsevier

tally obtained diffusion activation energy was then implemented in the homogeneous model. The pre-exponential factor was determined by visual correspondence with the experimental data. This resulted in an expression for the effective diffusion coefficient $D_{\text{eff}} = 3 \cdot 10^{-7} \cdot \exp(-\frac{43400}{RT})$. The obtained fits are included in Fig. 5.7. The numerical analysis exposed that the shape of the curves can be clarified by hydrogen diffusion processes only. An asymmetric shape arose for homogeneously charged samples, caused by the increasing diffusion upon increasing temperature. Similar shapes were experimentally observed when duplex stainless steels were homogeneously charged by gaseous charging [94, 95].

An average diffusion activation energy was determined based on the experimental data with a value in between diffusion in ferrite and austenite. Since the value is rather close to the one of austenite, it can be stated that most hydrogen diffused through several austenitic layers during leaving the sample. Hydrogen trapping in ferrite was not visible from the experimental results. Hydrogen atoms that are released from traps in ferrite are overruled by the large amount of hydrogen atoms that are released from austenite, i.e., the austenite diffusion peak overlaps with the peaks related to trapping in ferrite. Pu et al. [96] also argued that traps created through deformation of 304 stainless steel (dislocations and martensite) cannot be viewed in the large amount of hydrogen originating from interstitial positions in austenite. Evaluation of hydrogen trapping in duplex stainless steels appeared not to be possible as was shown by combining experimental results with diffusion models. Microstructural features with desorption activation energies greater than the diffusion activation energy of austenite can presumably be distinguished as a shoulder on TDS curves; however, they were not observed in the duplex stainless steel studied in this work. Hydrogen desorbing from austenite by diffusion determined completely the shape of the experimental TDS spectra. The main effect of the presence of the ferrite phase was an increase of the average diffusion kinetics.

5.3.2.2 Impact of Hydrogen on Active Deformation Mechanism of Duplex Stainless Steels

Figure 5.8 presents the EBSD measurements on the normal direction (ND) surface of the interrupted tensile tests in air (b), and for ex-situ (c) and in-situ (d) hydrogen-charged specimen. The initial condition, without hydrogen charging and deformation, is included as reference (a). Both image quality maps and phase maps are shown. Figure 5.9 displays a detailed EBSD measurement of a deformed austenite grain on the interrupted in-situ hydrogen-charged specimen. An image quality map, inverse pole figure (IPF) map and phase map are presented in this case. Moreover, several pole figures are included of specific crystallographic directions versus the sample reference system. These were constructed by first creating a partition based on crystal orientation with a maximal deviation of 5 °C. The orientations are indicated on the IPF map. High angle grain boundaries (>15 °C) are indicated in black in both figures.

Planar slip and both ε- and α'-martensite were detected in the austenite phase of the hydrogen-charged specimens being more noticeable for in-situ compared to ex-situ tensile testing. Both phenomena were absent for uncharged specimens. This could be clarified by a reduction in SFE, complemented by a change in what phase is accommodating most of the plastic deformation. Besides, additional pinning of edge dislocations by hydrogen atmospheres hindering cross-slip contributed to the difference in deformation mechanism. Hydrogen-induced martensitic transformations took place soon after yielding and were mainly characterized as ε-martensite. When higher strain levels were reached, hydrogen-assisted cracks initiated [97]. These HACs were studied with SEM-EBSD to identify the microstructurally most prone regions. Figure 5.10a shows an image of several cracks. The initiation characteristics were identified based on the phase map (cf. Fig. 5.10b) and ND IPF map (cf. Fig. 5.10c). The phase maps designated in which phase the crack initiated, whereas the ND IPF maps indicated whether the crack initiated on a grain boundary or inside a grain. Cracks, small enough to stay within one phase or boundary, were used for the statistic determination of the most favorable initiation site. Different categories

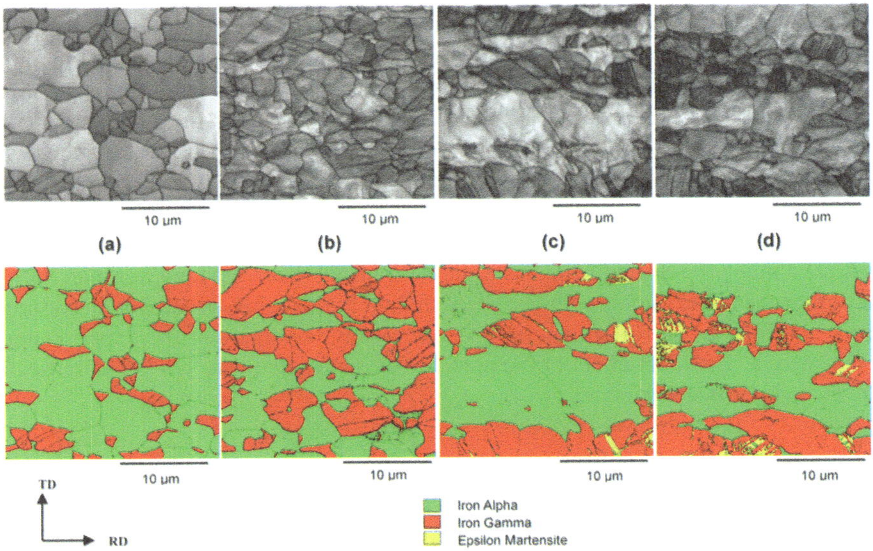

Fig. 5.8 EBSD measurements showing image quality and phase map on **a** initial state, **b** interrupted tensile test in air, **c** ex-situ interrupted tensile test and **d** in-situ interrupted tensile test [58] Copyright 2019, Elsevier

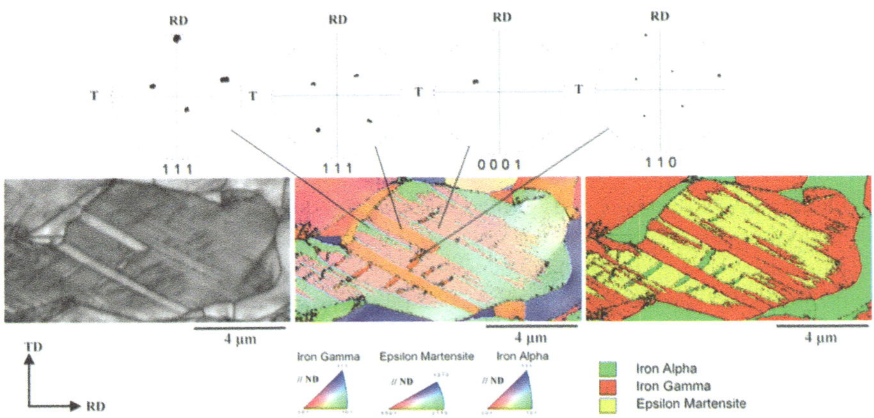

Fig. 5.9 Detailed EBSD measurement on hydrogen charged and in-situ deformed austenite grain. Various pole figures are included [58] Copyright 2019, Elsevier

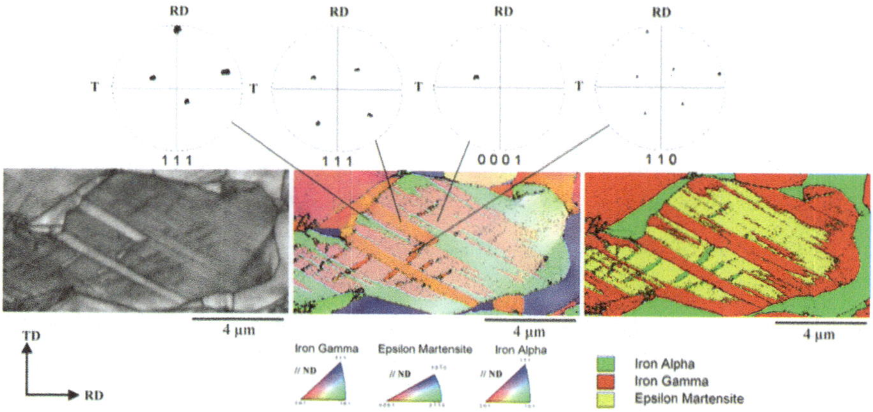

Fig. 5.10 EBSD characterization of HACs on ND plane after straining hydrogen-charged (1 day) duplex stainless steel to an intermediate elongation. **a** SEM image with type of crack (A = austenite, F = ferrite, M = martensite, IF = interface and GB = grain boundary), **b** phase map and (**c**) ND IPF map [97] Copyright 2020, Elsevier

were chosen; austenite grain interior, austenite/austenite grain boundary, austenite/martensite interface, ferrite grain interior or austenite/ferrite interface. The larger part (78.7%) of the cracks initiated in austenite (including the grain interior, grain boundaries of two adjacent austenitic grains and the austenite/martensite interface), followed by ferrite (16.6%) and finally the interface between austenite and ferrite (4.7%).

5.4 The Interaction of Hydrogen with Lab Cast Fe–C Alloys

In this section, a new in-situ bending technique is used to study the fracture mechanism of high-strength steels with limited ductility in the presence of hydrogen. The methodology was tested for generic Fe–C steels with a carbon content of 0.2, 0.4, and 1.1 wt.%. Figure 5.11a shows the hydrogen uptake capacity versus applied charging time, termed hydrogen saturation curves. The maximal amount of absorbed hydrogen increased with increasing carbon content. Fe–0.4C required longer charging times to reach hydrogen saturation than the Fe–0.2C due to a reduced hydrogen diffusivity. For the Fe–1.1C alloy, saturation is never reached. This can be attributed to the interaction between hydrogen with the present quench cracks in this material (cf. Fig. 5.1). Hydrogen charging caused additional cracking in this material [60]. However, for bending experiments with "saturated" samples, a charging time of two hours was chosen. No hydrogen-induced damage was observed for the other two materials. Figure 5.11b shows the bending curves of the Fe–0.4C samples. Although

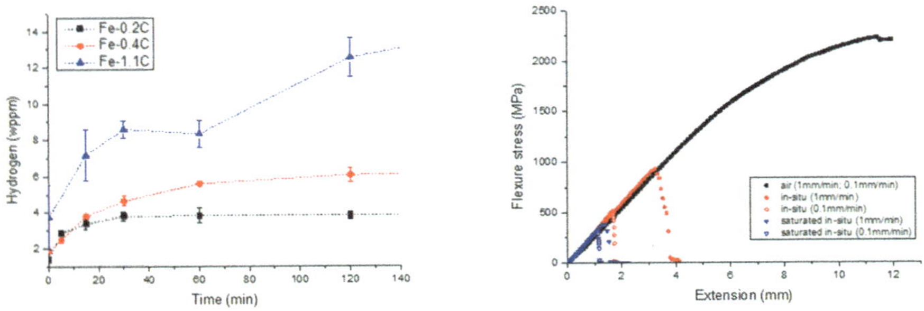

Fig. 5.11 Hydrogen saturation curves for the Fe–C alloys (**a**). Bending curves of Fe–0.4C, tested at 1 mm/min and 0.1 mm/min for different testing conditions (air, in-situ charged and in-situ charged after hydrogen pre-charging until saturation (cf. data shown in (**a**)) [60] Copyright 2020, Elsevier

Fe–0.4C samples are too brittle to be tested in tensile loading [27], they possess enough ductility for bending. In Fig. 5.11, the bending curves are only shown up to 12 mm of extension since from this point onward the support points change and there is a transition from three-point bending to four-point bending. Three different test conditions were compared; i.e., air, in-situ and saturated (pre-charged) in-situ, as described above. When the samples were charged with hydrogen, the flexure stress clearly decreased and fracture occurred abruptly. This effect is more pronounced when the samples were exposed to hydrogen for longer times, i.e., slower applied strain rates and/or with pre-charging. When the bending rate was decreased, hydrogen had more time to diffuse toward the highly stressed regions linked to the fracture initiation/propagation sites. When the samples were pre-charged, all the available trapping sites are occupied by hydrogen, and hydrogen can internally redistribute to the highly stressed parts and facilitate fracture. Due to these two factors, the sensitivity to failure increased for pre-charged samples and when lowering the bending deformation speed. The hydrogen embrittlement effect can be visualized by analyzing the related fracture surfaces by SEM. Figure 5.12 shows an overview of the appearance of the fracture surfaces for each testing condition.

In the absence of hydrogen, intergranular features and microvoids characterized the fracture surface. When the bending test was done with simultaneous hydrogen charging, additional cleavage facets, and cracks were found. When the in-situ test was done at a slower strain rate, the microvoids disappeared and the fracture surface was a mixture of intergranular and cleavage aspects accompanied with cracks. A similar fracture surface was detected when the sample was saturated before the test at 1 mm/min, although the cleavage fraction increased. Finally, when the sample was saturated and tested at lower strain rate, a transition to a pure cleavage fracture with dispersed cracks was seen. When hydrogen is added, the fracture mechanism changed since cracks perpendicular to the fracture surface were observed and the plastic strain marks almost completely disappeared. These perpendicular cracks run transgranular both along and through martensitic blocks (cf. Fig. 5.13b). The direction of the crack propagation changed with the orientation of the martensitic laths, indicating that

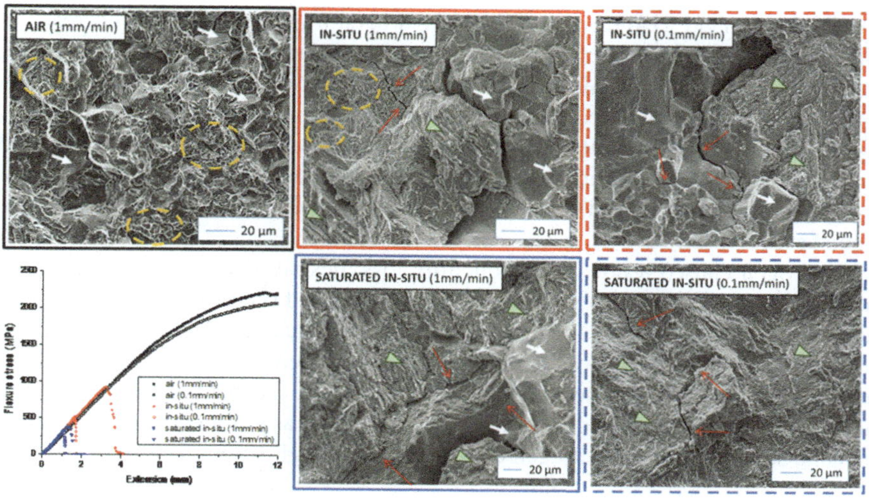

Fig. 5.12 SEM characterization of the fracture surfaces of Fe–0.4C for different conditions; air, in-situ, and saturated in-situ. Yellow dotted circles indicate zones with microvoids, white arrows intergranular features and green arrowheads cleavage features. Cracks are illustrated by fine red arrows [60] Copyright 2020, Elsevier

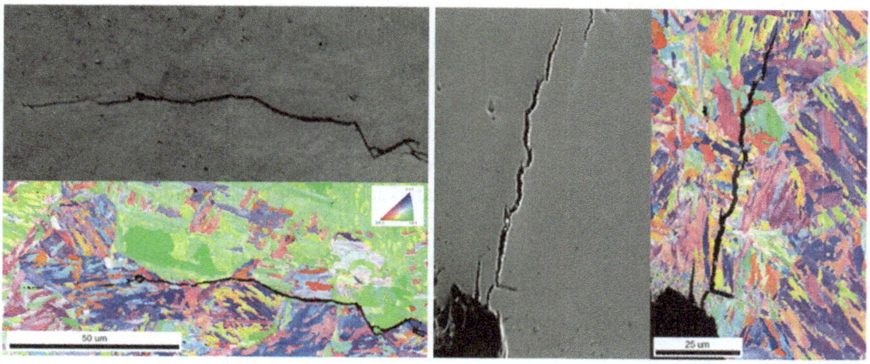

Fig. 5.13 EBSD scan of a crack running parallel to the fracture surface of Fe–0.4C subjected to bending in air (**a**), and a crack running perpendicular to the fracture surface in the presence of hydrogen (**b**) [60] Copyright 2020, Elsevier

cracks preferably ran along block boundaries. This is not the case in the air samples, where the crack propagated along the prior austenite grain boundaries (cf. Fig. 5.13a). Therefore, it can be stated that hydrogen weakened the block boundaries, which explained the shift to a cleavage type of fracture when hydrogen is present. As suggested by Nagao et al. [98], this quasi-cleavage fracture mode is a result of localized plasticity and a local hydrogen overconcentration at packet and/or block boundaries, which are high angle grain boundaries (HAGBs). On the one hand, HAGBs hinder dislocation movement causing a dislocation pile-up at the boundary.

On the other hand, HAGBs are effective hydrogen traps resulting in a decrease of the cohesive forces. These two phenomena weaken the block boundaries, causing quasi-cleavage features. This mechanism is termed "Hydrogen Enhanced Plasticity Mediated Decohesion". It can be considered as a mixture of HELP and HEDE, i.e., hydrogen activates different slip systems (HELP) which can intersect HAGBs and enhances hydrogen distribution by dislocations. This causes an increased amount of hydrogen deposited at the HAGBs which in turn weakens the cohesive forces at these boundaries (HEDE).

Hydrogen caused a noteworthy ductility loss, described by a transition from a microvoid (Fe–0.2C), intergranular (Fe–1.1C), or mixed (Fe–0.4C) fracture appearance to a cleavage type of fracture with additional cracking. The transition to the transgranular fracture type was explained by the Hydrogen Enhanced Plasticity Mediated Decohesion model. The similarities in the fracture mechanism for these alloys acknowledges the developed in-situ bending technique as a suitable method to evaluate the susceptibility to hydrogen embrittlement of high-strength steels.

5.5 Conclusion

This work provides an overview of our recent work on dedicated experimental methodologies to reveal specific aspects of the interaction of hydrogen with a steel microstructure. Firstly, the effect of constant tensile load on the hydrogen diffusivity was evaluated by electrochemical hydrogen permeation experiments during in-situ constant loading on DP steel. Secondly, a critical assessment of the evaluation of hydrogen trapping of low hydrogen diffusion materials was performed for duplex stainless steels. Moreover, the effect of hydrogen on the active deformation mechanism in this material was demonstrated as well by interrupted in-situ tensile tests combined with SEM-EBSD. Finally, the sensitivity of high strength-low ductility steels to hydrogen embrittlement was evaluated by in-situ bending tests. The operational embrittlement mechanism in these alloys was proposed based on SEM fractography. Key takeaways of this work are

1. Elastic applied tensile stresses increased the hydrogen diffusivity of DP steel due to the volume increase of the unit cell. When the imposed stress was equal to the yield stress, the increasing amount of hydrogen traps compensated for the increase of the diffusion coefficient resulting from the lattice expansion. Significant plastic deformation slowed down the diffusion due to the formation of lattice defects such as dislocations, vacancies, and vacancy clusters.
2. A combined experimental and numerical approach was used to increase the insights on the interpretation of TDS data of duplex stainless steel. It was verified that hydrogen desorbed through diffusion with an experimentally obtained value for the effective hydrogen diffusion coefficient in between the hydrogen diffusion coefficients in ferrite and austenite.
3. Planar slip and both ε- and α'-martensite were detected in the austenite phase of the hydrogen-charged duplex stainless steel specimens. This was more pronounced for in-situ compared to ex-situ tensile testing. Both phenomena were

absent in the uncharged condition. This was explained by a reduction in SFE together with a change in which phase accommodated most of the plastic deformation. Besides, additional pinning of edge dislocations by hydrogen atmospheres contributed to the large difference in observed deformation mechanism as well. At higher strain levels, hydrogen-assisted cracks initiated. Statistical analysis based on EBSD confirmed that the largest fraction of these cracks initiated in austenite.

4. The fracture mechanism in the presence of hydrogen for high strength-low ductility Fe–C steels was evaluated by in-situ bending. Hydrogen caused a noteworthy ductility loss, characterized by a transition from a microvoid (Fe–0.2C), intergranular (Fe–1.1C) or mixed (Fe–0.4C) fracture surface to a cleavage fracture surface with additional cracking. The transition to the transgranular fracture type is clarified by the Hydrogen Enhanced Plasticity Mediated Decohesion model, indicating that hydrogen was preferentially trapped at packet or block boundaries, causing a cleavage type of fracture.

Acknowledgements The authors wish to thank the senior postdoctoral fellowship of the Research Foundation-Flanders (FWO) via grant 12ZO420N and the Special Research Fund (BOF), UGent (grants BOF01P03516, BOF15/BAS/062 and BOF/01J06917) for support. Special thanks goes to Dr. E. Van den Eeckhout, L. Claeys and M. Pinson for their valuable contribution to the results presented in this chapter. The authors also acknowledge the technical staff from the Department Materials, Textiles and Chemical Engineering, UGent, for their help with the experiments and/or sample preparation.

References

1. Woodtli, J., Kieselbach, R.: Damage due to hydrogen embrittlement and stress corrosion cracking. Eng. Fail. Anal. **7**, 427–450 (2000)
2. Hilditch, T., Lee, S., Speer, J., Matlock, D.: Response to hydrogen charging in high strength automotive sheet steel products. SAE Technol. Paper (2003). https://doi.org/10.4271/2003-01-0525
3. Venezuela, J., Liu, Q., Zhang, M., Zhou, Q., Atrens, A.: The influence of hydrogen on the mechanical and fracture properties of some martensitic advanced high strength steels studied using the linearly increasing stress test. Corros. Sci. **99**, 98–117 (2015)
4. Depover, T., Escobar, D., Wallaert, E., Zermout, Z., Verbeken, K.: Effect of hydrogen charging on the mechanical properties of advanced high strength steels. Int. J. Hydrog. Energy **39**, 4647–4656 (2014)
5. Olden, V., Thaulow, C., Johnsen, R., Ostby, E., Berstad, T.: Influence of hydrogen from cathodic protection on the fracture susceptibility of 25%Cr duplex stainless steel - Constant load SENT testing and FE-modelling using hydorgen influenced cohersive zone elements. Eng. Fract. Mech. **76**, 827–844 (2009)
6. Johnson, W.: On some remarkable change produced in iron and steel by the action of hydrogen and acids. Proc. R. Soc. Lond. **23**, 168–179 (1875)
7. Hirth, J.: Effects of hydrogen on the properties of iron and steel. Metall. Trans. A **11A**, 861–890 (1980)
8. Oriani, R., Hirth, J., Smialowski, M.: Hydrogen Degradation of Ferrous Alloys. Noyes Publications, New Jersey (1985)

9. Sofronis, P.: Special issue on recent advances in engineering aspects of hydrogen embrittlement. Eng. Fract. Mech. **68**, 617–617 (2001)
10. Pundt, A., Kirchheim, R.: Hydrogen in metals: microstructural aspects. Ann. Rev. Mater. Res. **36**, 555–608 (2006)
11. Somerday, B., Sofronis, P., Jones, R. (eds.): Effects of Hydrogen on Materials. Wyoming, USA (2008)
12. Somerday, B., Sofronis, P., Jones, R. (eds.): Hydrogen-Materials Interactions. Wyoming, USA (2012)
13. Gangloff, P., Somerday, B.: Gaseous Hydrogen Embrittlement of Materials in Energy Technologies. Woodhead, Cambridge (2012)
14. Somerday, B., Gangloff, P., Jonas, R. (eds.): Effects of Hydrogen on Materials. Wyoming, USA (2016)
15. Revie, R., Sastri, V., Elboujdaii, M., Ramsingh, R., Lafrenière, Y.: Hydrogen-induced cracking of line pipe steels used in sour service. Corros. **49**, 531–536 (1993)
16. Maroef, I., Olson, D., Eberhart, M., Edwards, G.: Hydrogen trapping in ferritic steel weld metal. Int. Mater. Rev. **47**, 191–223 (2002)
17. Olden, V., Saai, A., Jemblie, L., Johnsen, R.: FE simulation of hydrogen diffusion in duplex stainless steel. Int. J. Hydrog. Energy **39**, 1156–1163 (2014)
18. Troiano, A.: The role of hydrogen and other interstitials in the mechanical behaviour of metals. Trans. ASM **52**, 54–80 (1960)
19. Beachem, C.: A new model for hydrogen-assisted cracking. Metall. Mater. Trans. B **3**, 441–455 (1972)
20. Lynch, S.: Comments on “A unified model of environment-assisted cracking”. Scr. Mater. **61**, 331–334 (2009)
21. Lynch, S.: Environmentally assisted cracking: overview of evidence for an adsorption-induced localised-slip process. Acta Metall. **36**, 2639–2661 (1988)
22. Nagumo, M.: Hydrogen related failure of steels—a new aspect. Mater. Sci. Technol. **20**, 940–950 (2004)
23. Gangloff, R.: Hydrogen assisted cracking of high strength alloys. In: Mine, I., Ritchie, R., Karihaloo, B. (eds.) Comprehensive Structural Integrity, pp. 31–101. Elsevier, Amsterdam (2003)
24. Barnoush, A., Vehoff, H.: Electrochemical nanoindentation: a new approach to prove hydrogen/deformation interaction. Scr. Mater. **55**, 195–198 (2006)
25. Birnbaum, H., Sofronis, P.: Hydrogen-enhanced localized plasticity—a mechanism for hydrogen-related fracture. Mater. Sci. Eng. A **176**, 191–202 (1994)
26. Depover, T., Wallaert, E., Verbeken, K.: Fractographic analysis of the role of hydrogen diffusion on the hydrogen embrittlement susceptibility of DP steel. Mater. Sci. Eng. A **649**, 201–208 (2016)
27. Depover, T., Wallaert, E., Verbeken, K.: On the synergy of diffusible hydrogen content and hydrogen diffusivity in the mechanical degradation o flaboratory cast Fe-C alloys. Mater. Sci. Eng. A **664**, 195–205 (2016)
28. Depover, T., Hertelé, S., Verbeken, K.: The effect of hydrostatic stress on the hydrogen induced mechanical degradation of dual phase steel: A combined experimental and numerical approach. Eng. Fract. Mech. **221**, 106704 (2019)
29. Depover, T., Verbeken, K.: The detrimental effect of mobile hydrogen at dislocations on the hydrogen embrittlement susceptibility of Fe-C-X alloys: an experimental proof of the HELP mechanism. Int. J. Hydrog. Energy **43**, 3050–3061 (2018)
30. Depover, T., Verbeken, K.: The effect of TiC on the hydrogen induced ductility loss and trapping behavior of Fe-C-Ti alloys. Corros. Sci. **112**, 308–326 (2016)
31. Kirchheim, R.: Bulk diffusion-controlled thermal desorption spectroscopy with examples for hydrogen in iron. Metall. Mater. Trans. A **47A**, 672–696 (2016)
32. Nagumo, M., Takai, K.: The predominant role of strain-induced vacancies in hydrogen embrittlement of steels: overview. Acta Mater. **165**, 722–733 (2019)

33. Kissinger, H.: Reaction kinetics in differential thermal analysis. Anal. Chem. **29**, 1702–1706 (1957)
34. Lee, J., Lee, J.: Hydrogen trapping in AISI-4340 steel. Metall. Sci. **17**, 426–432 (1983)
35. Lee, J., Lee, S.: Hydrogen trapping phenomena in metals with BCC and FCC crystal structures by the desorption thermal-analysis technique. Surf. Coat. Technol. **28**, 301–314 (1986)
36. Lee, S., Lee, J.: The trapping and transport phenomena of hydrogen in nickel. Metall. Trans. A **17A**, 181–187 (1986)
37. Verbeken, K.: Analysing hydrogen in metals: bulk thermal desorption spectroscopy (TDS) methods. In: Gangloff, P., Somerday, B. (eds.) Gaseous Hydrogen Embrittlement of Materials in Energy Technologies, pp. 27–55. Woodhead, Cambridge (2012)
38. Oriani, R.: The diffusion and trapping of hydrogen in steel. Acta Metall. **18**, 147–157 (1970)
39. Depover, T., Hajilou, T., Wan, D., Wang, D., Barnoush, A., Verbeken, K.: Assessment of the potential of hydrogen plasma charging as compared to conventional electrochemical hydrogen charging on dual phase steel. Mater. Sci. Eng. A **754**, 613–621 (2019)
40. Thomas, L., Li, D., Gangloff, R., Scully, J.: Trap-governed hydrogen diffusivity and uptake capacity in ultrahigh strength AERMET 100 steel. Metall. Mater. Trans. A **33**, 1991–2004 (2002)
41. Choo, W., Lee, J.: Thermal analysis of trapped hydrogen in pure iron. Metall. Mater. Trans. A **13**, 135–140 (1982)
42. Pressouyre, G.: Classification of hydrogen traps in steel. Metall. Trans. A **10A**, 1571–1573 (1979)
43. Di Stefano, D., Nazarov, R., Hickel, T., Neugebauer, J., Mrovec, M., Elsässer, C.: First principles investigation of hydrogen interaction with TiC precipitates in alpha-Fe. Phys. Rev. **93**, 184108–1–14 (2016)
44. Depover, T., Verbeken, K.: Evaluation of the effect of V4C3 precipitates on the hydrogen induced mechanical degradation in Fe-C-V alloys. Mater. Sci. Eng. A **675**, 299–313 (2016)
45. Drexler, A., Depover, T., Verbeken, K., Ecker, W.: Model-based interpretation of thermal desorption spectra of Fe-C-Ti alloys. J. Alloys Compd. **789**, 647–657 (2019)
46. Drexler, A., Depover, T., Leitner, S., Verbeken, K., Ecker, W.: Microstructural based hydrogen diffusion and trapping models applied to Fe-C-X alloys. J. Alloys Compd. **826**, 154057 (2020)
47. Wei, F., Hara, T., Tsuzaki, K.: Precise determination of the activation energy for desorption of hydrogen in two Ti-added steels by a single thermal-desorption spectrum. Metall. Mater. Trans B **35**, 587–597 (2004)
48. Depover, T., Verbeken, K.: Evaluation of the role of Mo2C in hydrogen induced ductility loss in Q&T Fe-C-Mo alloys. Int. J. Hydrog. Energy **41**, 14310–14329 (2016)
49. Sun, X., Xu, J., Li, Y.: Hydrogen permeation behaviour in austenitic stainless steel. Mater. Sci. Eng. A **114**, 179–187 (1989)
50. Laureys, A., Depover, T., Petrov, R., Verbeken, K.: Microstructural characterization of hydrogen induced cracking in TRIP-assisted steel by EBSD. Mater. Charact. **112**, 169–179 (2016)
51. Laureys, A., Pinson, M., Depover, T., Petrov, R., Verbeken, K.: EBSD characterization of hydrogen induced blisters and internal cracks in TRIP-assisted steel. Mater. Charact. **159**, 110029 (2020)
52. Laureys, A., Depover, T., Petrov, R., Verbeken, K.: Characterization of hydrogen induced cracking in TRIP-assisted steels. Int. J. Hydrog. Energy **40**, 16901–16912 (2015)
53. Depover, T., Wan, D., Wang, D., Barnoush, A., Verbeken, K.: The effect of hydrogen on the crack initiation site of TRIP-assisted steels during in-situ hydrogen plasma micro-tensile testing: Leading to an improved ductility. Mater. Charact. **167**, 110493 (2020)
54. Koyama, M., Tasan, C., Akiyama, E., Tsuzaki, K., Raabe, D.: Hydrogen-assisted decohesion and localized plasticity in dual-phase steel. Acta Mater. **70**, 174–187 (2014)
55. Liu, Q., Zhou, Q., Venezuela, J., Zhang, M., Atrens, A.: The role of the microstructure on the influence of hydrogen on some advanced high-strength steels. Mater. Sci. Eng. A **715**, 370–378 (2018)

56. Tasan, C., Diehl, M., Yan, D., Bechtold, M., Roters, F., Schemmann, L., Zheng, C., Peranio, N., Ponge, D., Koyama, M., Tsuzaki, K., Raabe, D.: An overview of dual phase steels: advances in microstructure-oriented processing and micromechanical guided design. Ann. Rev. Mater. Res. **45**, 391–431 (2015)
57. Kim, S., Yun, D., Jung, H., Kim, K.: Determination of hydrogen diffusion parameters of ferritic steel from electrochemical permeation measurement under tensile loads. J. Electrochem. Soc. **161**, 173–181 (2014)
58. Claeys, L., Depover, T., De Graeve, I., Verbeken, K.: First observation by EBSD of martensitic transformations due to hydrogen presence during straining of duplex stainless steel. Mater. Charact. **156**, 109843 (2019)
59. Claeys, L., Depover, T., De Graeve, I., Verbeken, K.: Electrochemical hydrogen charging of duplex stainless steel. Corros. **75**, 880–887 (2019)
60. Pinson, M., Springer, H., Depover, T., Verbeken, K.: Qualification of the in-situ bending technique towards the evaluation of the hydrogen induced fracture mechanism of martensitic Fe-C steels. Mater. Sci. Eng. A **792**, 139754 (2020)
61. Marder, A., Benscoter, A., Krauss, G.: Microcracking sensitivity in Fe-C plate martensite. Metall. Trans. **1**, 1545–1549 (1970)
62. Claeys, L., Cnockaert, V., Depover, T., De Graeve, I., Verbeken, K.: Critical assessment of the evaluation of thermal desorption spectroscopy data for duplex stainless steels: A combined experimental and numerical approach. Acta Mater. **186**, 1–9 (2020)
63. Crank, J.: The Mathematics of Diffusion. Clarendon Press, Oxford (1975)
64. Turnbull, A., Beylegaard, E., Hutchings, R.: Hydrogen transport in SAF 2205 and SAF 2507 duplex stainless steels. In: Turnbull, A. (ed.) Hydrogen Transport and Cracking in Metals, pp. 268–279. Woodhead, Cambridge (1994)
65. Devanathan, M., Stachurski, Z.: The adsorption and diffusion of electrolytic hydrogen in palladium. Proc. R. Soc. A **270**, 90–101 (1962)
66. Van den Eeckhout, E.: Methodological Study of the Hydrogen Diffusion in Steel by Means of Electrochemical Permeation. Ghent University, Ghent (2019)
67. Frappart, S.: Des éléments de compréhension sur les mécanismes de transport et de ségrégation de l'hydrogéne dans les aciers martensitiques trempés et revenus á haute limite d'élasticité. Université de La Rochelle, La Rochelle (2011)
68. Collet-Lacoste, J.: Intervention des phenomenes de surface sur la permeation electrochimique de l'hydrogene dans le fer: Role d'une couche d'oxyde ou de palladium. Université de Paris-Sud Centre d'Orsay, Paris (1993)
69. Pyun, S., Oriani, R.: The permeation of hydrogen through the passivating films on iron and nickel. Corros. Sci. **29**, 485–496 (1989)
70. Van den Eeckhout, E., Depover, T., Verbeken, K.: The effect of microstructural characteristics on the hydrogen permeation transient in quenched and tempered martensitic alloys. Metals **8**, 779 (2018)
71. Depover, T., Van den Eeckhout, E., Verbeken, K.: The impact of hydrogen on the ductility loss of bainitic Fe-C alloys. Mater. Sci. Technol. **32**, 1625–1631 (2016)
72. Van den Eeckhout, E., De Baere, I., Depover, T., Verbeken, K.: The effect of a constant tensile load on the hydrogen diffusivity in dual phase steel by electrochemical permeation experiments. Mater. Sci. Eng. A **773**, 138872 (2020)
73. Zhao, W., Zhang, T., He, Z., Sun, J., Wang, Y.: Determination of the critical plastic strain-induced stress of X80 steel through an electrochemical hydrogen permeation method. Electrochem. Acta **214**, 336–344 (2016)
74. Owczarek, E., Zakroczymski, T.: Hydrogen transport in a duplex stainless steel. Acta Mater. **48**, 3059–3070 (2000)
75. ASTM G148-97, Standard Practice for Evaluation of Hydrogen Uptake, Permeation, and Transport in Metals by an Electrochemical Technique (2011)
76. Frappart, S., Feaugas, X., Creus, J., Thebault, F., Delattre, L., Marchebois, H.: Study of the hydrogen diffusion and segregation into Fe-C-Mo martensitic HSLA steel using electrochemical permeation test. J. Phys. Chem. Solids **71**, 1467–1479 (2010)

77. Turnbull, A.: Perspectives on hydrogen uptake, diffusion and trapping. Int. J. Hydrog. Energy **40**, 16961–16970 (2015)
78. Namboodhiri, T., Nanis, L.: Concentration dependence of hydrogen diffusion in Armco iron. Acta Metall. **21**, 663–672 (1973)
79. Kumnick, A., Johnson, H.: Hydrogen transport through annealed and deformed Armco iron. Metall. Trans. **5**, 1199–1206 (1974)
80. Van den Eeckhout, E., Laureys, A., Van Ingelgem, Y., Verbeken, K.: Hydrogen permeation through deformed and heat-treated Armco pure iron. Mater. Sci. Technol. **33**, 1515–1523 (2017)
81. Essmann, U., Mughrabi, H.: Annihilation of dislocations during tensile and cyclic deformation and limits of dislocation densities. Philos. Mag. A **40**, 731–756 (1979)
82. Ohashi, T.: Generation and accumulation of atomic vacancies due to dislocation movement and pair annihilation. Philos. Mag. **98**, 2275–2295 (2018)
83. Doshida, T., Suzuki, H., Takai, K., Oshima, N., Hirade, T.: Enhanced lattice defect formation associated with hydrogen and hydrogen embrittlement under elastic stress of a tempered martensitic steel. ISIJ Int. **52**, 198–207 (2012)
84. Sakaki, K., Kawase, T., Hirato, M., Mizuno, M., Araki, H., Shirai, Y., Nagumo, M.: The effect of vacancy generation in iron by plastic deformation. Scr. Mater. **55**, 1031–1034 (2006)
85. Nagumo, M., Ohta, K., Saitoh, H.: Deformation induced defects in iron revealed by thermal desorption spectroscopy of tritium. Scr. Mater. **40**, 313–319 (1999)
86. Li, S., Li, Y., Lo, Y., Neeraj, T., Srinivasan, R., Ding, X., Sun, J., Qi, L., Gumbsch, P., Li, J.: The interaction of dislocation and hydrogen-vacancy complexes and its importance for deformation-induced proto nano-voids formatin in a-Fe. Int. J. Plast. **74**, 175–191 (2015)
87. Zhu, Y., Li, Z., Huang, M., Fan, H.: Study on interactions of an edge dislocation with vacancy-H complex by atomistic modelling. Int. J. Plast. **92**, 31–44 (2017)
88. Nagumo, M., Takai, K., Okuda, N.: Nature of hydrogen trapping sites in steels induced by plastic deformation. J. Alloys Compd. **293**, 310–316 (1999)
89. Huang, Y., Nakajima, A., Nishikata, A., Tsuru, T.: Effect of mechanical deformation on permeation of hydrogen in iron. ISIJ Int. **43**, 548–554 (2003)
90. Kim, S., Kim, K.: Electrochemical hydrogen permeation measurements through high-strength steel under uniaxial tensile stress in plastic range. Scr. Mater. **66**, 1069–1072 (2012)
91. Fallahmohammadi, E., Ballinger, R., Maruno, Y., Fumagalli, G., Bolzoni, F., Re, G., Lazzari, L.: Effect of plastic deformation on hydrogen diffusion of X65 pipeline steel. In: NACE International Conference (NACE-2014-4390), San Antonio (Texas), 9–13 March 2014
92. Luu, W., Liu, P., Wu, P.: Hydrogen transport and degradation of a commercial duplex stainless steels. Corros. Sci. **44**, 1783–1791 (2002)
93. Chen, S., Wu, T., Wu, J.: Effects of deformation on hydrogen degradation in a duplex stainless steel. J. Mater. Sci. **39**, 67–71 (2004)
94. Silverstein, R., Eliezer, D., Glam, B.: Hydrogen effect on duplex stainless steels at very high strain rates. Energy Proc. **107**, 199–204 (2017)
95. Park, Y., Maroef, I., Landau, A., Olson, D.: Retained austenite as a hydrogen trap in steel welds. Welding Res. **81**, 27–35 (2002)
96. Pu, S., Turk, A., Lenka, S., Ooi, S.: Study of hydrogen release resulting from the transformation of austenite into martensite. Mater. Sci. Eng. A **754**, 628–635 (2019)
97. Claeys, L.De., Graeve, I., Depover, T., Verbeken, K.: Hydrogen-assisted cracking in 2205 duplex stainless steel: Initiation, propagation and interaction with deformation-induced martensite. Mater. Sci. Eng. A **797**, 140079 (2020)
98. Nagao, A., Dadfarnia, M., Somerday, B., Sofronis, P., O. Ritchie, R.: Hydrogen-enhanced-plasticity mediated decohesion for hydrogen-induced intergranular and "quasi-cleavage" fracture of lath martensitic steels. J. Mech. Phys. Solids **112**, 403–430 (2018)

Chapter 6
Modeling of Hydrogen Diffusion in Slow Strain Rate (SSR) Testing of Notched Samples

Andreas Drexler, Josef Domitner, and Christof Sommitsch

Abstract Macroscopic testing of the hydrogen embrittlement (HE) resistivity of ultra and advanced high-strength steels is still a difficult task. Different testing procedures are recommended in literature, such as the slow strain rate (SSR) test, the constant load (CL) test, or the incremental step load (ISL) test. Nevertheless, a direct comparison of the results of the different testing procedures is challenging and the influence of the microstructure is not well understood. Therefore, the present work contributes to a deeper understanding of the role of internal hydrogen diffusion and trapping at microstructural defects during SSR testing of notched samples using physical reasonable diffusion-mechanical finite element (FE) simulations. The modeling approach allows a detailed study of the role of macroscopic strength and multiple trapping sites on the local hydrogen accumulation at the notch.

Keywords Hydrogen diffusion · Beneficial trapping · Hydrogen embrittlement · Slow strain rate testing · Notch effect

6.1 List of Symbols

A	mm^2	Sample surface area
c	wppm	Total hydrogen concentration
c_{im}	wppm	Immobile hydrogen concentration
c_L	wppm	Lattice hydrogen concentration
$c_{L,0}$	wppm	Lattice hydrogen concentration under zero hydrostatic stress
$c_{T,i}$	wppm	Trapped hydrogen concentration
D_{chem}	mm^2/s	Chemical diffusion coefficient
D_L	mm^2/s	Tracer diffusion coefficient

A. Drexler (✉) · J. Domitner · C. Sommitsch
Institute of Materials Science, Joining and Forming (IMAT), Graz University of Technology (TU Graz), Inffeldgasse 11/I, 8010 Graz, Austria
e-mail: andreas.drexler@tugraz.at

V. A. Polyanskiy and A. K. Belyaev (eds.), *Advances in Hydrogen Embrittlement Study*, Advanced Structured Materials 143, https://doi.org/10.1007/978-3-030-66948-5_6

E	GPa	Young's modulus
$E_{B,i}$	kJ/mol	Binding energy of traps
$E_{B,deep}$	kJ/mol	Binding energy of deep trapping sites of 60 kJ/mol
$E_{B,shallow}$	kJ/mol	Binding energy of shallow trapping sites of 30 kJ/mol
E_S	kJ/mol	Solution enthalpy
ε_{eng}	–	Engineering strain
ε_{pl}	–	Equivalent plastic strain
ε_{eqv}	–	Equivalent strain
γ	mm^{-2}	Correlation factor between dislocation density and equivalent plastic strain
J	mol/mm^2s	Total hydrogen flux
$J_{\nabla c}$	mol/mm^2s	Chemical hydrogen flux
$J_{\nabla\sigma}$	mol/mm^2s	Stress-driven hydrogen flux
K	MPa	Strength coefficient
K_i	–	Equilibrium constant of trapping sites i
$K_{\nabla\sigma}$	–	Equilibrium constant for volumetric strained interstitial lattice sites
k	mol/mm	Number of shallow trap sites per unit dislocation length
L_0	mm	Length of the unstrained gauge section of a sample
L	mm	Length of the strained gauge section of a sample
l	mm	Size of the plastic zone
m	$wppm/MPa^{1/2}$	Constant factor of Sievert's law
M_H	g/mol	Molar mass of hydrogen
μ_L	kJ/mol	Chemical potential of the lattice hydrogen in bulk
$\mu_{P_{H_2}}$	kJ/mol	Chemical potential of the gaseous hydrogen molecules
N	–	Total number of trapping sites
N_L	mol/mm^3	Density of interstitial lattice sites
$N_{T,0}$	mol/mm^3	Initial trap density
$N_{T,i}$	mol/mm^3	Trap density of trapping sites i
n	–	Strain hardening exponent
ν	–	Poisson constant
p_{H_2}	MPa	Hydrogen partial pressure
R_g	J/molK	Universal gas constant
ρ	g/mm^3	Density
σ_H	MPa	Hydrostatic stress
σ_{Mises}	MPa	Von Mises stress
σ_y	MPa	Yield stress
$\sigma_{y,0}$	MPa	Initial yield stress
T	K	Temperature
t	s	Hydrogen charging time
V	mm^3	Sample volume
V_H	mm^3/mol	Partial molar volume of hydrogen
$y_{T,i}$	–	Site fraction of trapping sites i
y_L	–	Lattice site fraction

6.2 Introduction

Hydrogen in metals, especially in advanced high-strength steels, can cause premature macroscopic brittle failure [1]. Different test procedures for dog bone-shaped samples are available to investigate the susceptibility of the microstructure to hydrogen embrittlement ranging from constant load (CL) tests to slow strain rate (SSR) tests. The incremental step load (ISL) tests combine aspects of both SSR and CL tests. Comparing the susceptibility of different materials to hydrogen embrittlement (HE) or testing under real hydrogen donating conditions coined the terms "internal" hydrogen embrittlement (IHE), hydrogen "environmental" embrittlement (HEE), and hydrogen-induced stress corrosion cracking (SCC). Lynch [2] has already stated that the hydrogen source can trigger different embrittlement sites within the microstructure. It may change the rate-controlled process from hydrogen diffusion to hydrogen adsorption by controlling the kinetics of IHE and HEE. The distinction between IHE, HEE, or SCC makes sense from an experimental macroscopic point of view, however, the microscopic HE mechanisms [3, 5–7] do not distinguish different hydrogen sources. Evaluation of the material susceptibility to hydrogen embrittlement is often done by comparing time to failure and threshold stresses, which are measured with CL or SSR tests. However, both quantities are also influenced by other factors than microstructure, like

1. hydrogen charging procedure,
2. sample geometry,
3. strain rate, and
4. applied mechanical load.

This indicates a strong dependency of the time to failure and threshold stress on hydrogen absorption, diffusion, and hydrogen accumulation in critical zones [8]. Therefore, the roles of internal hydrogen diffusion and trapping in SSR testing of hydrogen pre-charged and notched samples must be investigated. For that purpose, a physically reasonable, fully coupled diffusion-mechanical finite element (FE) model [9] was developed and parametrized. Using this FE model, the roles of

1. competitive mechanisms of stress-driven hydrogen diffusion and shallow hydrogen trapping by means of local plastic deformation,
2. "beneficial" deep trapping sites,
3. strain rate dependency,
4. mechanical strength, and
5. strain hardening

on the local hydrogen accumulation at the notch are studied in detail.

6.3 Multiphysical Modeling of Slow Strain Rate (SSR) Tests

It is well known that hydrostatic stresses and plastic strain have a crucial effect on the macroscopic hydrogen accumulation at the notch [10]. For that purpose, the fully coupled diffusion-mechanical model of Barrera et al. [9] was implemented in the FE software package Abaqus [11] using the UMATHT subroutine. Barrera has shown how to compute directly the gradient of hydrostatic stress field inside the UMATHT using common block variables with USDFLD and GETVRM subroutines and how to consider hydrogen trapping on the overall chemical diffusion [10, 12].

6.3.1 Sample Geometry and Finite Element Meshing

A notched 1-mm thick sample was studied with a notch radius of 0.1 mm and a notch angle of 60°, as shown in Fig. 6.1a. The initial length of the gauge section was 60 mm. Due to symmetry conditions, only a quarter of the sample was meshed with 8-node plane stress diffusionally coupled quadrilateral elements (CPS8T). The minimum mesh size was 1 μm directly at the notch root, while the maximum mesh size was 0.5 mm at the top side of the sample, as shown in Fig. 6.1b.

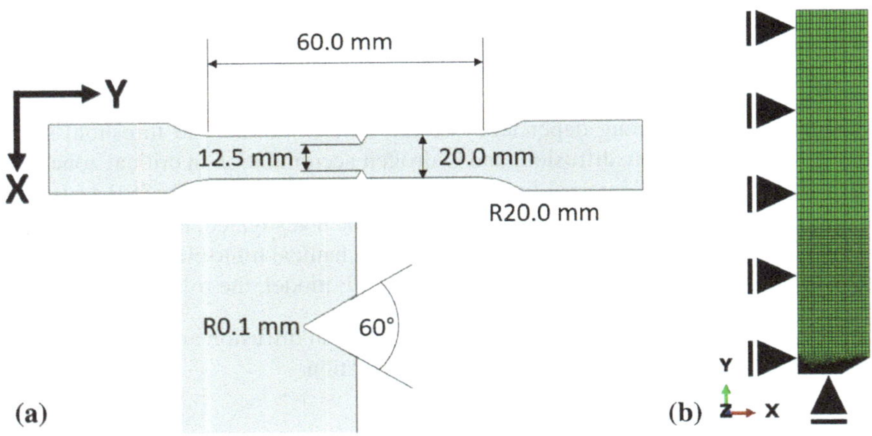

Fig. 6.1 **a** Sketch of the notched sample considered in the multiphysical FE model. **b** Due to symmetry conditions, only a quarter of the sample was modeled

6.3.2 Mechanical Material Modeling

Linear elastic-plastic material behavior with isotropic material hardening was considered. The yield condition was expressed as

$$\sigma_{\mathrm{Mises}} = \sigma_y, \tag{6.1}$$

where σ_{Mises} is the von Mises stress and σ_{y} is the yield stress. The evolution of the yield stress with plastic deformation is defined as the sum of the initial yield stress $\sigma_{\mathrm{y,0}}$ and the hardening function, according to Hollomon and Ludwik,

$$\sigma_{\mathrm{y}} = \sigma_{\mathrm{y,0}} + K\varepsilon_{\mathrm{pl}}^{n}\,, \tag{6.2}$$

where K is the strength coefficient and n is the strain hardening exponent. $\varepsilon_{\mathrm{pl}}$ is the equivalent plastic strain. Neither a strain rate dependency of the material hardening, as observed in advanced high-strength steels at low temperatures [13], nor softening [9] with increasing hydrogen concentration was considered in the present study. Furthermore, the presented model does not consider the influence of damage by means of micropores or microcracks on the hydrogen accumulation during SSR testing [8, 14].

6.3.3 Diffusional Modeling of Lattice Hydrogen

The total hydrogen flux J is the sum of the chemical hydrogen flux $J_{\nabla c}$ driven by the hydrogen concentration gradient according to Fick's first law and the stress-driven hydrogen flux $J_{\nabla\sigma}$ [15]:

$$J = J_{\nabla\mathrm{c}} + J_{\nabla\sigma} = -D_{\mathrm{L}}\nabla c_{\mathrm{L}} + \frac{D_{\mathrm{L}} c_{\mathrm{L}} V_{\mathrm{H}}}{R_{\mathrm{g}} T}\nabla\sigma_{\mathrm{H}}. \tag{6.3}$$

In Eq. (6.3), D_{L} is the tracer diffusion coefficient of the interstitial lattice sites, c_{L} is the lattice hydrogen concentration, V_{H} is the partial molar volume of hydrogen, R_{g} is the universal gas constant, and T is the temperature. The hydrostatic stress is defined as $\sigma_{\mathrm{H}} = \frac{1}{3}(\sigma_{11} + \sigma_{22} + \sigma_{33})$. Pipe diffusion [16] of hydrogen, e.g., along dislocation networks, or grain boundary diffusion was not considered in the diffusion model, because of its negligible influence in bcc iron [17]. This assumption was also confirmed by Siegl et al. [18] studying the influence of the grain size and plastic deformation on the chemical hydrogen diffusion in bcc iron for industrial relevant grain sizes of (10 μm and larger).

6.3.4 Hydrogen Micro-segregation Modeling

Mass conservation requires that the total hydrogen flux through the surfaces of an element is equal to the change of total hydrogen concentration within this element. The total hydrogen concentration c is the sum of the interstitial lattice hydrogen concentration c_L, the reversible trapped hydrogen concentration $c_{T,i}$ at microstructural defects i and the immobile hydrogen concentration c_{im}:

$$c = c_L + \sum_{i=1,N} c_{T,i} + c_{im}. \tag{6.4}$$

N is the total number of reversible trapping sites within the microstructure. Figure 6.2 illustrates the hydrogen energy states of the microstructure with respect to the concentrations of lattice hydrogen, reversible trapped hydrogen, and immobile hydrogen.

Lattice hydrogen occupies interstitial lattice positions, such as tetragonal sites in bcc iron [19, 20], which provides fast diffusion paths between neighboring reversible trapping sites. In ferrite activation energies of lattice hydrogen diffusion from 5 kJ/mol to 10 kJ/mol [21–24] are reported, which enable hydrogen exchange between neighboring reversible trapping sites (mean free distance of approximately 1 nm) within nanoseconds or less. The direct exchange between lattice hydrogen and reversible trapped hydrogen can be expressed in the following reaction equation as

$$H_L + V_T \leftrightharpoons V_L + H_T. \tag{6.5}$$

H_L are interstitial lattice hydrogen atoms, V_T are vacant reversible trapping sites, V_L are vacant interstitial lattice sites, and H_T are reversibly trapped hydrogen atoms. Hydrogen trapping at microstructural defects does not alter the microstructure and can be regarded as a reversible reaction. According to Toribio and Kharin [25]

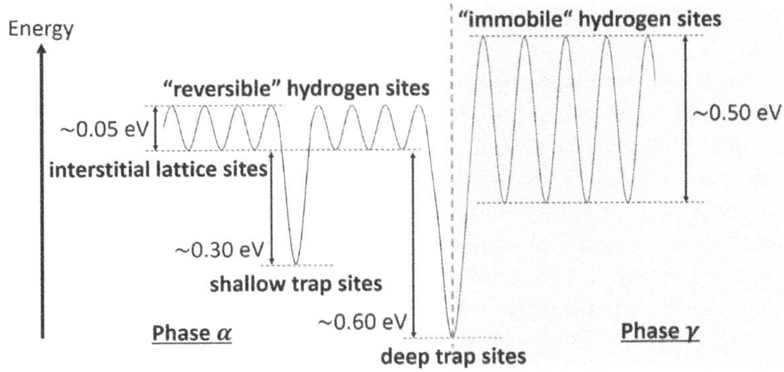

Fig. 6.2 Schematic illustration of the hydrogen energy states of the microstructure considered in this work

equilibrium between lattice hydrogen and trapped hydrogen is locally reached, if the time for long-range hydrogen diffusion is much larger than the characteristic time for local relaxation between interstitial lattice hydrogen and trapped hydrogen. Applying the generalized Oriani's [26] equation relates the site fraction of reversible trapped hydrogen $y_{\mathrm{T,i}} = c_{\mathrm{T,i}}/N_{\mathrm{T,i}}$ to the site fraction of lattice hydrogen $y_{\mathrm{L}} = c_{\mathrm{L}}/N_{\mathrm{L}}$:

$$\frac{y_{\mathrm{L}}(1 - y_{\mathrm{T,i}})}{y_{\mathrm{T,i}}(1 - y_{\mathrm{L}})} = \exp\left(-\frac{E_{\mathrm{B,i}}}{R_{\mathrm{g}}T}\right) = K_{\mathrm{i}}. \tag{6.6}$$

N_{L} is the density of interstitial lattice sites and $N_{\mathrm{T,i}}$ is the trap density. K_{i} is the equilibrium constant and $E_{\mathrm{B,i}}$ is the binding energy of the reversible trapping sites i. Reversible trapping sites provide a variety of binding energies ranging up to approximately 100 kJ/mol. Immobile hydrogen differs from reversible trapped hydrogen according to its expansion in the microstructure and to its activation energy for local hydrogen diffusion. The concept is new to the authors' knowledge and differs from the often cited theory of "quasi-irreversible" trapped hydrogen [27]. Immobile hydrogen occupies interstitial lattice sites in secondary phases with very low diffusion rates at room temperature, e.g., retained austenite or titanium carbides [28, 29]. The activation energy for local hydrogen diffusion in retained austenite or in carbides is reported as 50 kJ/mol or higher [29, 30]. Phase transformation or thermal activation at elevated temperatures release the immobile hydrogen to the matrix.

6.3.5 Constitutive Modeling of Trap Density Evolution

Microstructural defect densities and the density of reversible hydrogen trapping sites $N_{\mathrm{T,i}}$ evolve during industrial thermomechanical processing of a material:

1. Heat treatments at elevated temperatures [31–34] cause growth and coarsening of carbides, which change the number of available trapping sites at the interfaces between carbides and matrix [28, 35].
2. Cold forming is the main factor at room temperature changing the reversible trap densities in the material by increasing vacancy and dislocation densities [17, 36].

A further constitutive equation for the evolution of the trap density N_{T} with equivalent plastic deformation is [17]

$$N_{\mathrm{T}} = N_{\mathrm{T,0}} + k\gamma\varepsilon_{\mathrm{pl}}, \tag{6.7}$$

with $N_{\mathrm{T,0}}$ being the initial trap density and k and γ being semi-empirical material constants. Figure 6.3 compares Eq. (6.7) with respect to other evolution equations suggested in literature [10, 17, 37–39]. The trap densities introduced by plastic deformation cover a range of 10^{-13}–10^{-5} mol/mm^3. The evolution equations and material constants derived by Song et al. [37] and Drexler et al. [17] agree very well.

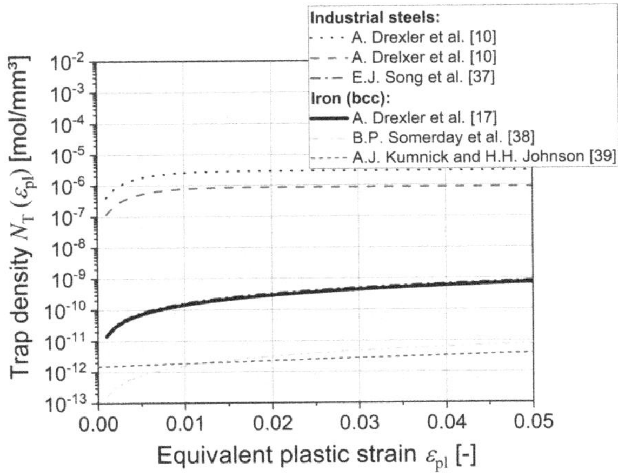

Fig. 6.3 The hydrogen trap density increases with increasing equivalent plastic strain

6.4 Model Parametrization

To apply the diffusion-mechanical FE model for SSR testing, mechanical and diffusional model parameters have to be defined. The identification of mechanical model parameters is well established in literature and can be based on classical tensile tests. The parametrization of Eq. (6.2) allows to extrapolate the stress–strain behavior also to larger equivalent plastic strain. In this study, all simulation results are referenced to the yield curve with $\sigma_{y,0} = 1000$ MPa, $K = 1200$ MPa and $n = 0.5$. This "reference" yield curve is indicated by a bold black line in Fig. 6.4a, b. To study the influence of the initial yield stress on the SSR tests, $\sigma_{y,0}$ was increased from 600 MPa to 1400 MPa with $K = 1200$ MPa and $n = 0.5$, as shown in Fig. 6.4a. To study the effect of strain hardening on the local hydrogen accumulation at the notch, K was increased from 0 MPa (ideal plastic deformation) to 1900 MPa, as shown in Fig. 6.4b, while $\sigma_{y,0} = 1000$ MPa and $n = 0.5$.

The identification of the diffusional model parameters is a difficult task, especially for a broad application range. The diffusional material parameters of the lattice hydrogen (D_L, N_L) were measured by electrochemical permeation on well-annealed bcc iron samples. Appropriate permeation samples need a minimum thickness of 1–2 mm and palladium coatings, especially on the oxidation sides [18, 21] to reduce the influence of surface reactions and hydrogen recombination. Assuming that the samples were nearly trap-free allows to determine the tracer hydrogen diffusion coefficient D_L and the density of the interstitial lattice sites N_L. A previous literature survey revealed $D_L = 9 \cdot 10^{-3}$ mm²/s for well-annealed iron at room temperature [40]. This value was confirmed by Siegl et al. [18]. The results from electrochemical permeation measurements, especially the activation energy of lattice hydrogen diffusion, are in good agreement with DFT calculations [41] representing ideal defect-free iron crystals. The trapping parameters, as the total number of trapping sites N, and the corresponding binding energies $E_{B,i}$ and trap densities $N_{T,i}$, can be identi-

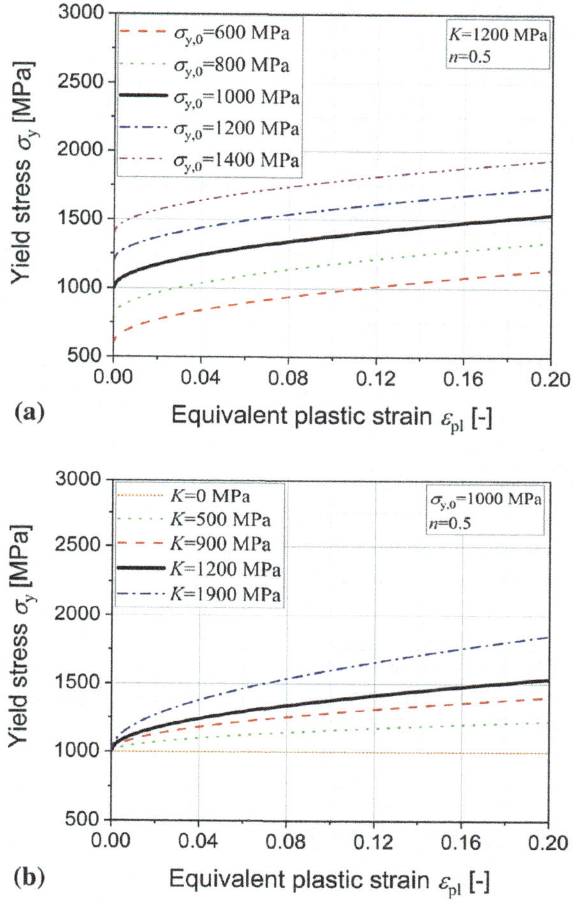

Fig. 6.4 Yield stress as function of equivalent plastic strain according to Eq. (6.2) and corresponding model parameters. While in (**a**) only the initial yield stress $\sigma_{y,0}$ is changed, in (**b**) the strength coefficient K is changed. The strain hardening exponent n remained constant at 0.5

fied by means of thermal desorption spectroscopy (TDS) [10, 28, 42]. An effective "two-trap" approach was applied in the present study considering shallow and deep trapping sites. The binding energies of the shallow and deep trapping sites, respectively, assumed to be $E_{B,shallow} = 30$ kJ/mol and $E_{B,deep} = 60$ kJ/mol. The shallow trapping sites represent mixed dislocations or subgrain boundaries (misorientation less than 15°). Deep trapping sites can be introduced into the microstructure by vacancies or carbide interfaces. Note that carbides not only add deep trapping sites, they also add high densities of shallow trapping sites. The initial trap densities for the shallow and deep trapping sites are both assumed to be $N_{T,0} = 10^{-10}$ mol/mm^3. Only the trap density of the shallow trapping sites evolves during plastic deformation at room temperature according to Eq. (6.7). Due to the high binding energy of the deep trapping sites, the equilibrium constant is approximately unity and all reversible deep trapping sites are fully occupied by hydrogen at room temperature.

Table 6.1 Summary of the model parameters. The parameters for the stress–strain curves can be found in Fig. 6.4

Parameter	Symbol	Value	Uni	References
Tracer diffusion coefficient	D_{L}	$9 \cdot 10^{-3}$	mm^2/s	[24, 43]
Shallow binding energy	$E_{\mathrm{B,shallow}}$	30	kJ/mol	
Deep binding energy	$E_{\mathrm{B,deep}}$	60	kJ/mol	
Young's modulus of iron	E	210	GPa	
Correlation factor	κ	$2 \cdot 10^9$	mm^{-2}	[44]
Number of trap sites per unit length of dislocations	k	$7.2 \cdot 10^{-18}$	mol/mm	[17]
Molar mass of hydrogen	M_{H}	1.008	g/mol	
Density of interstitial lattice sites	N_{L}	$2 \cdot 10^{-4}$	mol/mm^3	[15]
Shallow trap density	$N_{\mathrm{T,shallow}}$	10^{-10}	mol/mm^3	
Deep trap density	$N_{\mathrm{T,deep}}$	10^{-10}	mol/mm^3	
Poisson constant of iron	ν	0.3	–	
Density or iron	ρ	$7.9 \cdot 10^{-3}$	g/mm^3	
Partial molar volume	V_{H}	$1.2 \cdot 10^{-3}$	mm^3/mol	[45]

Therefore, increasing the binding energy of the deep trapping sites would not change the simulation results. Finally, all mechanical and diffusional model parameters are summarized in Table 6.1.

6.5 Results and Discussion

6.5.1 Modeling of Gaseous and Electrochemical Pre-charging

An initial total hydrogen concentration of $c = 0.2$ wppm is considered in the FE simulations. This low concentration should represent, e.g., corrosive hydrogen uptake of advanced high-strength steels during service. The initial hydrogen concentration c is homogenously distributed throughout the sample. During the simulations of the SSR tests, hydrogen accumulates at the notch and depletes at the notch flanks. To study the role of "beneficial" deep trapping sites on the hydrogen accumulation during SSR testing, it is assumed that the initial total hydrogen concentration increases during gaseous hydrogen pre-charging and remains unchanged in electrochemically pre-charged samples. Gaseous hydrogen charging occurs under constant outer hydrogen partial pressure at elevated temperatures. The temperature is needed to dissociate the hydrogen molecules and to increase the chemical hydrogen adsorption and the absorption rates. As illustrated in Fig. 6.5a, the outer partial pressure and temperature define the applied chemical potential. In equilibrium, the chemical potentials μ_L and $\mu_{P_{H_2}}$ of the lattice hydrogen in bulk and of the gaseous hydrogen molecules, respectively, are equal at the surface [46]. Thus, the equilibrium lattice hydrogen concentration c_L depends only on the square root of the outer hydrogen partial pressure $\sqrt{p_{H_2}}$, which is known as Sieverts' law, as

$$c_L = m\sqrt{p_{H_2}}\exp\left(-\frac{E_S}{RT}\right) = \text{const.}, \tag{6.8}$$

with p_{H_2}, E_S and m being the outer hydrogen partial pressure, the solution energy and a constant factor, respectively. This leads to an increase of the initial total hydrogen concentration c by adding deep trapping sites to the microstructure.

Figure 6.5b illustrates the hydrogen uptake via the surface during electrochemical hydrogen charging. It is a complex chemical and electrochemical process. Different chemical and electrochemical reactions including Volmer, Tafel, Heyrovsky, and absorption reactions occur at the surface. These reactions depend strongly on hydrogen charging parameters, such as temperature, electrolyte velocity, thickness of the boundary layer, sample geometry, applied current density, oxide layers, or hydrogen flux to the bulk. In stationary state, the electron transfer via the surface and the concentration of the chemically adsorbed hydrogen H_{ad} are constant. Electrochemical permeation experiments are suitable to estimate the stationary hydrogen absorption flux J, which is approximately 5–10% of the applied current density [18] for a given sample thickness. In a first approximation electrochemical pre-charging can be modeled by assuming constant hydrogen surface flux $J = \text{const.}$ [47, 48]. This leads to a defined total hydrogen concentration in the samples after a charging time t and does not change by adding deep trap sites, as

$$c = JA\frac{t}{V} = \text{const.}, \tag{6.9}$$

where A is the sample surface and V is the sample volume.

6.5.2 *Local Model-Based Evaluation of Notched Tensile Samples*

It is well known that the strain rate dependency of SSR tests depends on different parameters, as the notch geometry or the chemical hydrogen diffusion. Due to localized plastic deformation, the local equivalent strain ε_{eqv} at the notch root differs from the engineering strain ε_{eng} in tension direction:

$$\varepsilon_{eng} = \frac{L - L_0}{L_0}, \tag{6.10}$$

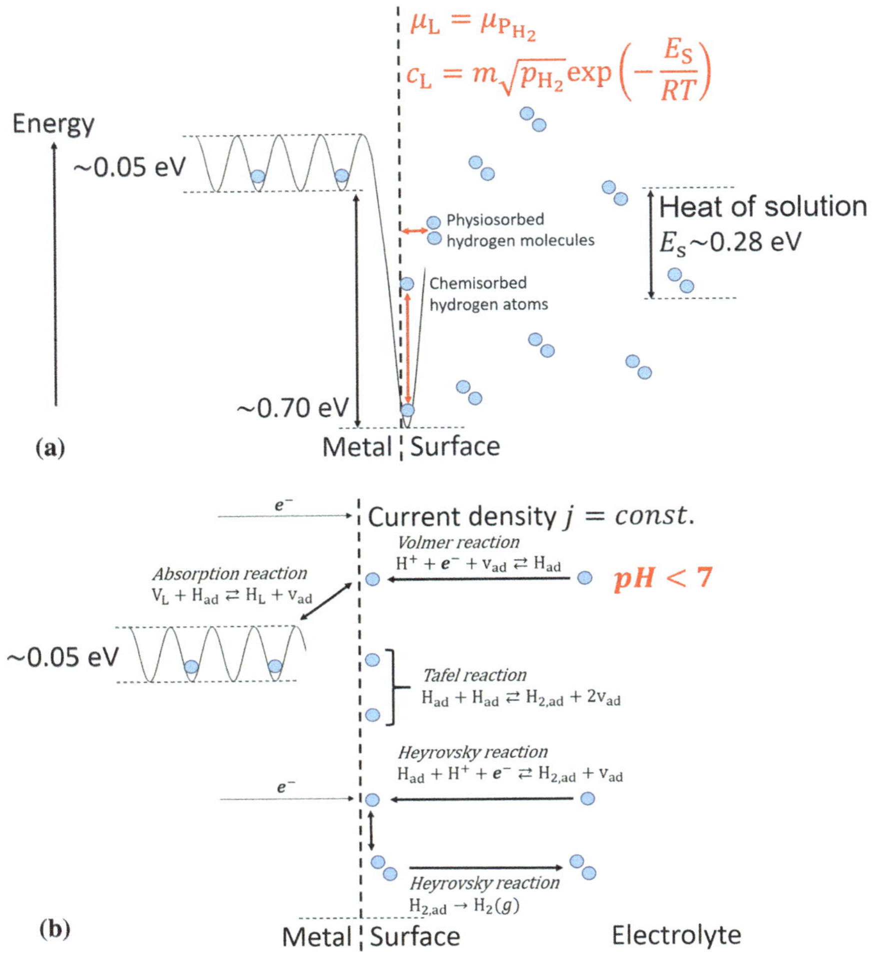

Fig. 6.5 **a** Illustration of the gaseous hydrogen uptake at the metal surface. **b** Illustration of the electrochemical hydrogen uptake at the metal surface in acidic electrolytes (pH < 7)

with $L_0 = 30$ mm being the length of the unstrained gauge section and L being the current length of the gauge section. The equivalent strain ε_{eqv} was evaluated according to von Mises. Figure 6.6 correlates the local equivalent strain at the notch root with the engineering strain ε_{eng}. The local equivalent strain increases faster than the engineering strain. In the linear elastic regime of the sample, the local equivalent strain is about 9 times larger than the engineering strain. With the onset of local plastic deformation, the local equivalent strain increases even faster at the notch root. In other words, the local equivalent strain in the plastically and volumetrically strained zone is a magnitude larger than the engineering strain. The same is true

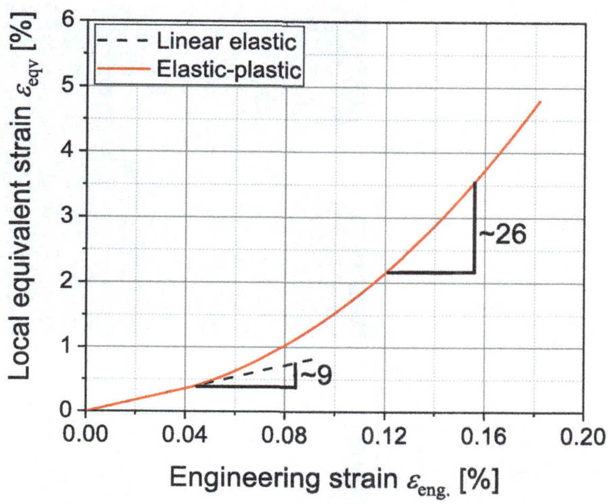

Fig. 6.6 Local equivalent strain at the notch root as function of applied engineering strain for linear elastic and elastic-plastic material behaviors

for the local strain rates. This difference between local and engineering quantities is very important and illustrates the influence of the sample geometry on the SSR test results.

Figure 6.7 shows the distributions of the equivalent plastic strain ε_{pl} and of the hydrostatic stress σ_H , respectively, in the notched sample for an engineering strain of $\varepsilon_{eng} = 0.17$ %. Due to the high initial yield stress of $\sigma_{y,0} = 1000$ MPa with $K = 1200$ MPa and $n = 0.5$, the plastic deformation localizes at the notch. This effect is even more pronounced with increasing yield stress. The zone of the hydrostatic stress rise at the notch is circular shaped. Due to the long-range interactions between hydrogen and volumetrically strained interstitial lattice sites, the hydrostatic stress peak at the notch can even attract hydrogen from the notch flanks and from the surface.

Both the localized equivalent plastic strain ε_{pl} and the hydrostatic stress σ_H influence the hydrogen redistribution during SSR testing of notched and hydrogen precharged samples. According to Eq. (6.7), the trap density increases linearly with increasing equivalent plastic strain at the notch, as shown in Fig. 6.8a. Trapping at the various lattice defects created by equivalent plastic strain and stress concentration causes local hydrogen accumulation with a maximum at the notch root. Due to the isolated surface conditions, e.g., by means of galvanization after hydrogen precharging, hydrogen redistributes at the notch, as shown in Fig. 6.8b. As expected, most of the hydrogen is provided from the notch flanks and from the surface areas. To understand the strain rate dependency of the hydrogen accumulation due to hydrogen diffusion, the theoretical concept of chemical diffusion must be considered. The chemical diffusion coefficient D_{chem} links the chemical hydrogen flux $J_{\nabla c}$ with the gradient of the total hydrogen concentration ∇c [49] instead of the gradient of the

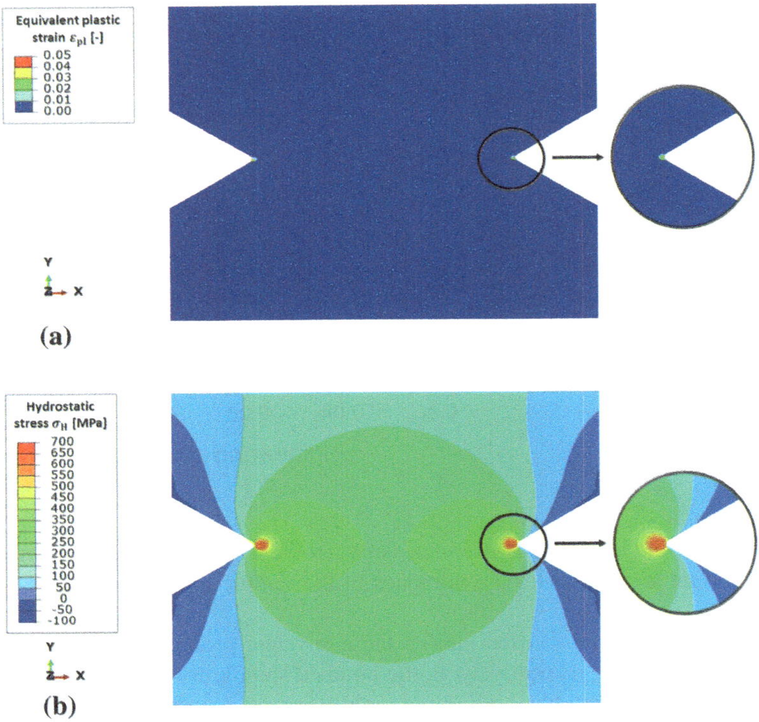

Fig. 6.7 **a** Equivalent plastic strain and **b** hydrostatic stress in the notched sample for an engineering strain of 0.17 % and a corresponding nominal stress of 600 MPa

lattice hydrogen concentration ∇c_L, as in Eq. (6.3). The chemical diffusion coefficient is defined as

$$D_{chem} = D_L \frac{\partial c_L}{\partial c}. \tag{6.11}$$

In general, D_{chem} differs from the tracer diffusion coefficient D_L, as used in Eq. (6.3), and from the effective diffusion coefficient D_{eff} [15, 28, 35, 50]. Due to the thermodynamic factor $\frac{\partial c_L}{\partial c}$, the chemical hydrogen diffusion coefficient depends on the local hydrogen concentration and on the trap density and it varies in the sample, especially in the plastically and volumetrically strained zone at the notch root. D_{chem} increases with increasing hydrogen concentration c of the reversible hydrogen trapping sites, decreases with temperature and decreases with plastic deformation. A decrease of D_{chem} would retard the hydrogen accumulation and most likely increases the strain rate dependency of the SSR tests. As shown in Fig. 6.8c, only at a very small area at the notch root D_{chem} decreases about 10%.

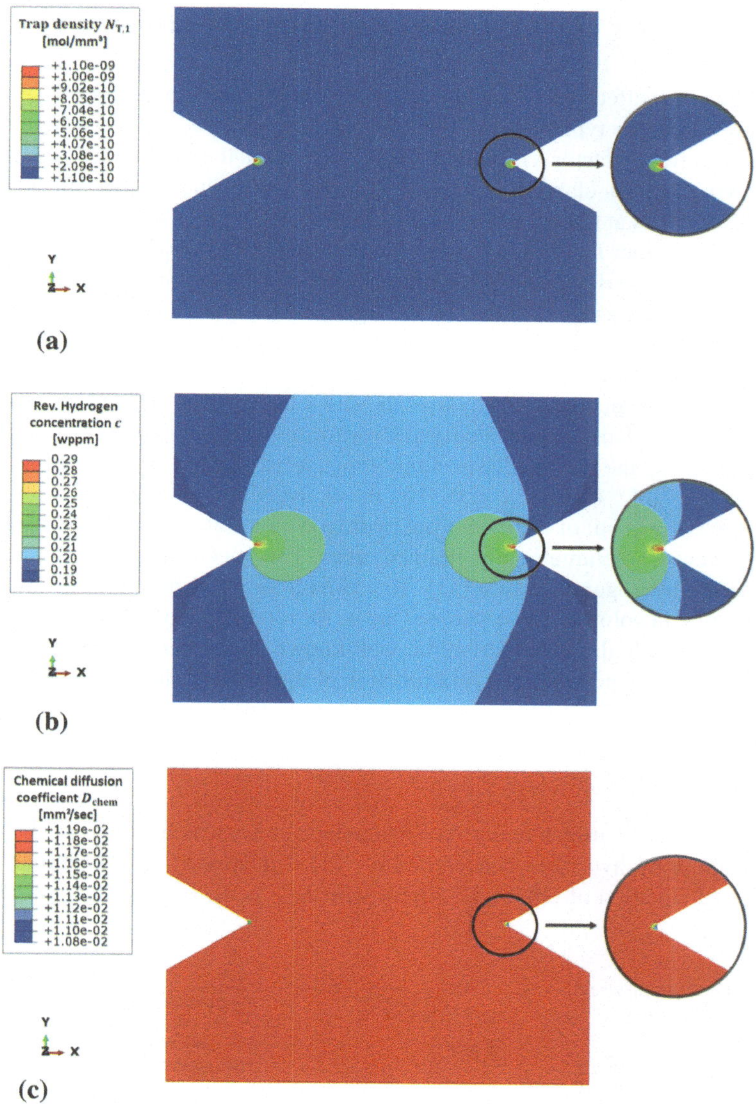

Fig. 6.8 **a** Trap density of reversible shallow trapping sites, **b** reversible total hydrogen concentration and **c** chemical diffusion coefficient D_{chem} in the notched sample for the engineering strain of 0.17%, the strain rate of 10^{-8} 1/s and the initial homogenously distributed total hydrogen concentration of $c = 0.2$ wppm

6.5.3 Influence of Hydrostatic Stress and Hydrogen Trapping

Due to the notch effect, stress concentration occurs at the notch. The degree of stress concentration under typically tension loads can be expressed as a non-dimensional stress concentration factor K_t. Mechanical FE analysis of the notched sample revealed $K_t = 7.5$ in the linear elastic regime. Due to the stress concentration, localized plastic deformation starts at the notch root at an engineering stress of $\sigma_{eng} = 130$ MPa and spreads with further increasing tension along the ligament into bulk. This results in the shift of the hydrostatic stress maximum of few micrometers. Figure 6.9 compares the hydrostatic stress profile along the half ligament of the notched sample for a linear elastic and an elastic-plastic material model. Linear elastic simulation revealed the maximum of hydrostatic stress directly at the surface, while elastic-plastic material behavior shifts the maximum of hydrostatic stress from the surface towards the bulk. To compare simulations with identical hydrostatic stress of around 650 MPa, the linear elastic and the elastic–plastic simulations were stopped at engineering strains of $\varepsilon_{eng} = 8.6 \cdot 10^{-2}\%$ and $\varepsilon_{eng} = 0.17\%$, respectively.

Owing to the misfit of the interstitial hydrogen atoms at nanoscale, lattice hydrogen segregates in volumetrically strained areas. This mechanism is well known as stress-driven hydrogen diffusion [51]. To estimate the equilibrium lattice hydrogen concentration in volumetrically strained areas, the total hydrogen flux, Eq. (6.3), was set to zero, which allows to derive the well-known exponential expression for the lattice hydrogen concentration c_L as function of the hydrostatic stress [10]:

$$\frac{c_L}{c_{L,0}} = \exp\left(-\frac{V_H \sigma_H}{R_g T}\right) = K_{\sigma_H}. \tag{6.12}$$

$c_{L,0}$ is the lattice hydrogen under zero hydrostatic stress which should not be mixed up with the total hydrogen concentration c [1]. Equation (6.12) is similar to the generalized equation of Oriani, Eq. (6.6), with K_{σ_H} as the equilibrium constant for

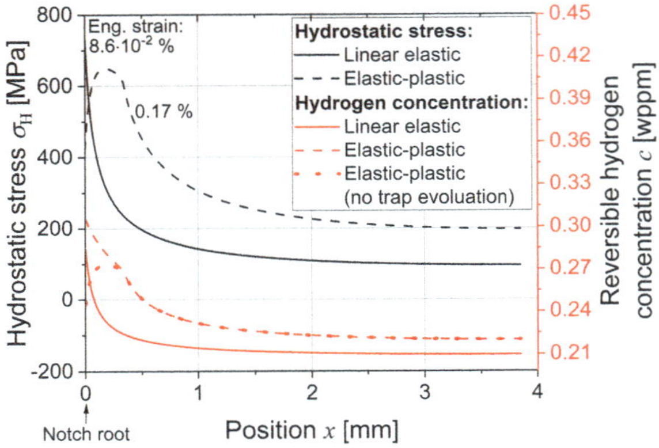

Fig. 6.9 Hydrostatic stress in the ligament of the notched sample and corresponding equilibrium reversible trapped hydrogen concentration according to Eq. (6.12)

volumetrically strained interstitial lattice sites. Comparing $V_H\sigma_H$ with the binding energy E_B of hydrogen trapping sites at microstructural defects gives a corresponding energy of approximately about 1 kJ/mol for a hydrostatic stress of 65 MPa. In addition to the local accumulation of lattice hydrogen, reversible hydrogen trapping reinforces the hydrogen accumulation. Applying Eq. (6.12) together with Eq. (6.6), the equilibrium total reversible hydrogen concentration c can be estimated along the ligament using the simulated hydrostatic stress curve, as shown in Fig. 6.9. The calculations reveal the maximum reversible total hydrogen concentration at the notch root. Neglecting the interaction between plastic deformation and shallow hydrogen trapping, as indicated by the bold dashed line in Fig. 6.9, would shift the maximum of total reversible hydrogen concentration from the surface to the bulk. In this case, the maximum total reversible hydrogen concentration would coincident with the position of the maximum hydrostatic stress. In other words, the localized equivalent plastic deformation and stress concentration at the notch are crucial for understanding hydrogen embrittlement phenomenon on the macroscale.

6.5.4 Influence of Yield Stress, Strength, and Strain Hardening

It is well known that the susceptibility of metallic alloys to hydrogen embrittlement [52, 53] increases with increasing strength [54]. In high-strength steels, less than 1 wppm of measured total hydrogen concentration can provoke time-delayed macroscopic brittle failure of structural components. Hydrogen concentrations are always measured on bulk specimens, however, local hydrogen concentrations, e.g., at notches or at severe deformed edges [42], can be 5–10 times higher than the bulk concentrations [42, 55, 56]. In other words, conventional hydrogen analysis techniques cannot distinguish between homogeneous and inhomogeneous hydrogen distributions. Using diffusion-mechanical FE models allows to study the macroscopic influence of the yield stress and of the strain hardening on the local hydrogen accumulation in hydrogen pre-charged and notched samples. Due to the stress concentration and stress triaxiality, the equivalent plastic strain localizes at the notch and reaches higher values than in unnotched tensile samples at equal engineering strains. To study the influence of the initial yield stress on the local hydrogen accumulation, $\sigma_{y,0}$ is increased from 600 MPa to 1400 MPa. The strain hardening was unchanged with $K = 1200$ MPa and $n = 0.5$, as shown in Fig. 6.4a. Figure 6.10 shows the accumulation of total reversible hydrogen concentration as function of initial yield stress and engineering strain rate at an equivalent plastic strain of 3%. Hence, the trap density at the evaluation node is equal in all simulations. With increasing yield stress, the hydrogen accumulation is reduced at the notch root. With decreasing engineering strain rate, the hydrogen accumulation increases and reaches a quasi-static state below 10^{-7} s^{-1}. The local reversible hydrogen concentration increases from an initially homogeneous distributed hydrogen concentration of 0.2 wppm to

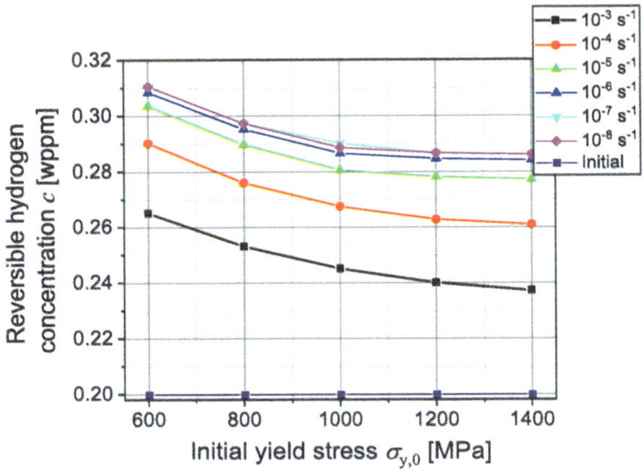

Fig. 6.10 Effect of initial yield stress and strain rate on the local reversible hydrogen concentration at an equivalent plastic strain of 3% at the notch root

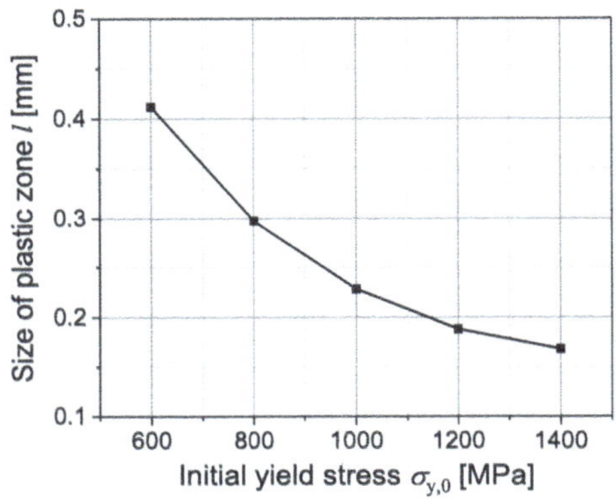

Fig. 6.11 Effect of initial yield stress on plastic localization at the notch root. The size of the plastic zone is given for constant equivalent plastic strain of 3% at the notch root

around 0.31 wppm at $\sigma_{y,0} = 600$ MPa and to around 0.28 at $\sigma_{y,0} = 1400$ MPa. Nevertheless, the influence of the yield stress seems to be more pronounced below $\sigma_{y,0} = 1000$ MPa.

With increasing yield stress, the size of the plastic zone decreases and the maximum of the hydrostatic stress increases. According to Fig. 6.11, the plastic deformation localizes at the notch with increasing yield stress. Comparing simulations at the given equivalent plastic strain of 3%, the size of the plastic zone decreases from 0.41 to 0.18 mm. Due to the higher yield stress and smaller plastic zone, the maximum of the hydrostatic stress increases from 132 MPa to 195 MPa at the maximum equivalent plastic strain of 3%, and the peak position of hydrostatic stress was closer to the surface. In other words, the decrease of the accumulated hydrogen concentration with increasing yield stress can be attributed to the smaller size of the plastic zone and the increasing influence of the stress-driven hydrogen flux pointing from the surface towards bulk.

To evaluate the influence of strain hardening on the local hydrogen accumulation, K is increased from 0 MPa (ideal plasticity) to 1900 MPa. The initial yield stress

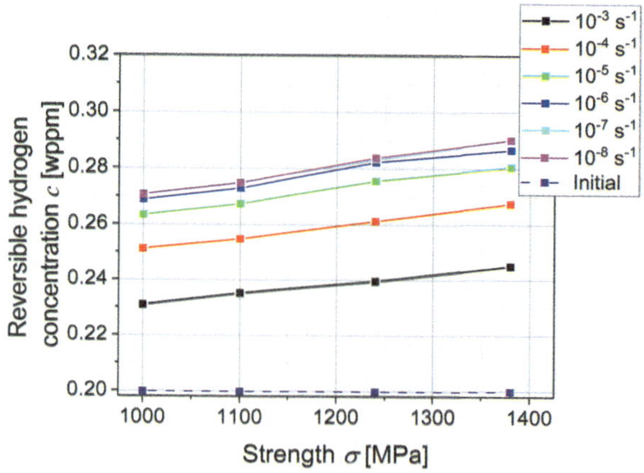

Fig. 6.12 Effect of the strength and strain rate on the local reversible hydrogen concentration at an equivalent plastic strain of 4% at the notch root

is $\sigma_{y,0} = 1000$ MPa, and the strain hardening exponent is $n = 0.5$. The corresponding stress-equivalent plastic strain curves are summarized in Fig. 6.4b. As shown in Fig. 6.12, with increasing strain hardening and thus with increasing strength, more hydrogen accumulates at the notch. The local hydrogen concentrations and strength were evaluated at a local equivalent plastic strain of 4%. Hence, the local shallow trap density at the surface is equal in all simulations. The size of the plastic zone increases slightly from 0.2 to 0.3 mm with increasing strength. The maximum of hydrostatic stress increases with increasing size of plastic deformation from 144 to 191 MPa. However, the increase of the hydrogen concentration at the surface mainly results from the size increase of the plastic zone, which shifts the maximum hydrostatic stress further towards the bulk.

6.5.5 Influence of "beneficial" Deep Trapping Sites

Besides the conventional hydrogen solubility in the interstitial lattice sites, atomic hydrogen can trap at foreign substitutional atoms, dislocations, grain boundaries, carbides, microcracks, or micropores [51]. The concept of "beneficial" trapping is very controversial. Only those deep trapping sites, which do not initiate cracks, are beneficial in the sense of hydrogen embrittlement, since strongly trapped hydrogen does not participate in nanoscale hydrogen embrittlement mechanisms. The nature of "beneficial" deep trapping sites is still unexplored, and, e.g., carbides or grain boundaries may increase both deep and shallow trap densities at the same time. To study the "beneficial" effect of deep trapping sites on the hydrogen accumulation at the notch during SSR testing, three cases are compared:

1. Case 1: only shallow trapped hydrogen with a total reversible hydrogen concentration of $c = 0.2$ wppm and a corresponding initial lattice hydrogen concentration of $c_{L,0} = 1.5 \cdot 10^{-9}$ mol/mm^3.
2. Case 2: shallow and deep trapped hydrogen with an initial lattice hydrogen concentration of $c_{L,0} = 1.5 \cdot 10^{-9}$ mol/mm^3 and a total reversible hydrogen concentration of $c = 0.213$ wppm.
3. Case 3: shallow and deep trapped hydrogen with a total hydrogen concentration of $c = 0.2$ wppm and an initial lattice hydrogen concentration of $c_{L,0} = 1.4 \cdot 10^{-9}$ mol/mm^3.

Figures 6.13 and 6.14 show the accumulation of the total reversible hydrogen concentration c at the notch root as function of

1. the engineering strain rate and
2. the equivalent plastic strain.

With decreasing engineering strain rate, hydrogen has more time to diffuse and accumulate locally at the notch. Quasi-static conditions are reached for strain rates below 10^{-7} s^{-1} and the local reversible hydrogen concentration depends only on the equivalent plastic strain. With increasing equivalent plastic strain, the total reversible hydrogen concentration increases. For quasi-static conditions and equivalent plastic strain of 5% at the notch root, the total reversible hydrogen concentration increased from $c = 0.2$ wppm to $c = 0.29$ wppm. Figures 6.13 and 6.14 show the local accumulation of total reversible hydrogen concentration at the notch root with respect to

1. multiple trapping,
2. engineering strain rate,
3. equivalent plastic strain, and
4. hydrogen pre-charging procedure.

While Fig. 6.13 considers gaseous hydrogen pre-charging with $c_{L,0} = 1.5 \cdot 10^{-9}$ mol/mm^3, Fig. 6.14 considers electrochemical hydrogen pre-charging with $c = 0.2$ wppm. With respect to Fig. 6.13 more reversible hydrogen accumulates in the plastically and volumetrically strained zone of gaseous pre-charged samples, but the additional deep trapping sites do not affect the strain rate dependency. The difference to case 1 is exactly 0.013 wppm. Applying electrochemical pre-charging, as shown in Fig. 6.14, causes a decrease of the hydrogen concentration at the notch root compared to gaseous hydrogen pre-charging. This difference is small, but increased slightly with decreasing strain rates. In general, the site fraction $y_{T,\text{deep}}$ of the deep trapping sites with the binding energy of $E_{B,\text{deep}} = 60$ kJ/mol is almost unity at room temperature. Therefore, gaseous hydrogen pre-charging increases the total reversible hydrogen concentration by $\Delta c = 0.013$ wppm compared to case 1. In addition, the fully occupied deep trapping sits cannot contribute to the chemical diffusion coefficient and remains the engineering strain rate dependency unchanged. Applying electrochemical pre-charging prescribes the total reversible hydrogen concentration.

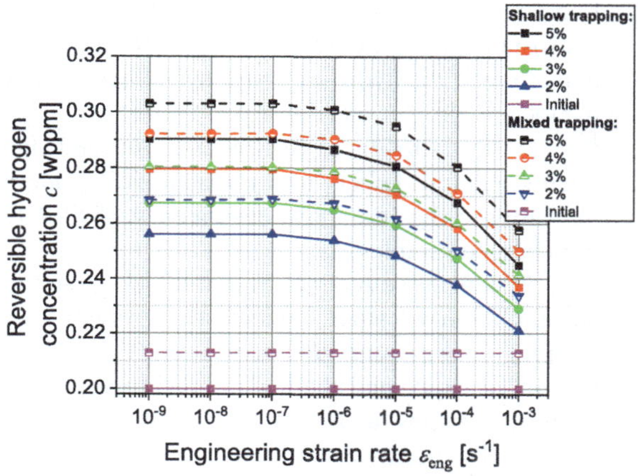

Fig. 6.13 Hydrogen concentration at the notch root as function of engineering strain rate and trapping: shallow hydrogen trapping with binding energy $E_{B,shallow} = 30$ kJ/mol, mixed hydrogen trapping with binding energies $E_{B,shallow} = 30$ kJ/mol and $E_{B,deep} = 60$ kJ/mol and an initial total hydrogen concentration $c = 0.213$ wppm

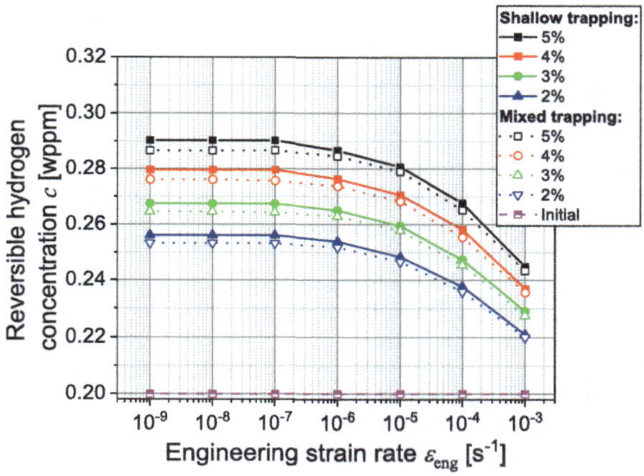

Fig. 6.14 Hydrogen concentration at the notch root as function of engineering strain rate and trapping: shallow hydrogen trapping with binding energy $E_{B,shallow} = 30$ kJ/mol, mixed hydrogen trapping with binding energies $E_{B,shallow} = 30$ kJ/mol and $E_{B,deep} = 60$ kJ/mol and an initial total hydrogen concentration $c = 0.2$ wppm

Therefore, adding deep trapping sites to the microstructure reduces the amount of lattice and shallow trapped hydrogen. Especially, the lower lattice concentration reduces both the stress-driven hydrogen flux and the chemical diffusion coefficient. In other words, the lower lattice hydrogen concentration affects the strain rate dependency during SSR testing, although, the deep trapping sites are fully occupied.

6.6 Conclusion

The following conclusions can be drawn from diffusion-mechanical simulations of slow strain rate (SSR) testing of hydrogen pre-charged and notched samples:

1. The notch effect leads to localized plastic deformation and stress concentration, and increases the local strain rates compared to the global strain rates.

2. Localized plastic deformation during SSR testing shifts the peak position of the hydrostatic stress along the ligament from surface towards bulk, decreases the maximum of the hydrostatic stress and increases the dislocation density and thus the local hydrogen trap density.
3. Due to stress-driven hydrogen diffusion and due to hydrogen segregation to microstructural defects in the plastic zone, internal hydrogen contributes significantly to the strain rate dependency of SSR tests of notched and hydrogen pre-charged samples.
4. The "beneficial" effect of reversible deep trapping sites on the local hydrogen accumulation was studied with respect to strain rate, plastic deformation, multiple trapping, and pre-charging conditions. Deep trapping sites can only reduce the amount of lattice and shallow trapped hydrogen, if hydrogen uptake during pre-charging remains unaffected. Especially, the effect of deep trapping sites on the interstitial lattice hydrogen concentration needs to be further investigated.

References

1. Nagumo, M.: Fundamentals of Hydrogen Embrittlement. Springer Singapore, Singapore (2016)
2. Lynch, S.: Discussion of some recent literature on hydrogen-embrittlement mechanisms: addressing common misunderstandings. Corros. Rev. **37**, 377–395 (2019). https://doi.org/10.1515/corrrev-2019-0017
3. Beachem, C.D.: A new model for hydrogen-assisted cracking (hydrogen "embrittlement"). Metall. Mater. Trans. B. **3**, 441–455 (1972). https://doi.org/10.1007/BF02642048
4. Birnbaum, H.K., Sofronis, P.: Hydrogen-enhanced localized plasticity-a mechanism for hydrogen-related fracture. Mater. Sci. Eng. A. **176**, 191–202 (1994). https://doi.org/10.1016/0921-5093(94)90975-X
5. Troiano, A.R.: The role of hydrogen and other interstitials in the mechanical behavior of metals. Metallogr. Microstruct. Anal. **5**, 557–569 (2016). https://doi.org/10.1007/s13632-016-0319-4
6. Oriani, R.A.: Whitney Award lecture-1987: hydrogen-the Versatile Embrittler. Corrosion **43**, 390–397 (1987). https://doi.org/10.5006/1.3583875
7. Lynch, S.P.: Progress towards understanding mechanisms of hydrogen embrittlement and stress corrosion cracking. NACE Int. Corros. Conf. Ser. **074931–0749355** (2007)
8. Schaffner, T.: Charakterisierung und Modellierung des Wasserstofftransports und der wasserstoffabhängigen Beanspruchungsgrenzen höchstfester Mehrphasenstähle (2018)
9. Barrera, O., Tarleton, E., Tang, H.W., Cocks, A.C.F.: Modelling the coupling between hydrogen diffusion and the mechanical behaviour of metals. Comput. Mater. Sci. **122**, 219–228 (2016). https://doi.org/10.1016/j.commatsci.2016.05.030
10. Drexler, A., Bergmann, C., Manke, G., Kokotin, V., Mraczek, K., Pohl, M., Ecker, W.: On the local evaluation of the hydrogen susceptibility of cold-formed and heat treated advanced high strength steel (AHSS) sheets. Mater. Sci. Eng. A. **800**, 140276 (2021). https://doi.org/10.1016/j.msea.2020.140276
11. Dassault Systèmes Simulia: Abaqus Unified FEA. https://www.3ds.com/products-services/simulia/products/abaqus/
12. Massone, A., Manhard, A., Jacob, W., Drexler, A., Ecker, W., Hohenwarter, A., Wurster, S., Kiener, D.: An SEM compatible plasma cell for in situ studies of hydrogen-material interaction. Rev. Sci. Instrum. **91**, 043705 (2020). https://doi.org/10.1063/1.5142043

13. Cao, Y., Karlsson, B., Ahlström, J.: Temperature and strain rate effects on the mechanical behavior of dual phase steel. Mater. Sci. Eng. A. **636**, 124–132 (2015). https://doi.org/10.1016/j.msea.2015.03.019
14. Koyama, M., Tasan, C.C., Akiyama, E., Tsuzaki, K., Raabe, D.: Hydrogen-assisted decohesion and localized plasticity in dual-phase steel. Acta Mater. **70**, 174–187 (2014). https://doi.org/10.1016/j.actamat.2014.01.048
15. Svoboda, J., Fischer, F.D.: Modelling for hydrogen diffusion in metals with traps revisited. Acta Mater. **60**, 1211–1220 (2012). https://doi.org/10.1016/j.actamat.2011.11.025
16. Fischer, F.D., Svoboda, J., Kozeschnik, E.: Interstitial diffusion in systems with multiple sorts of traps. Model. Simul. Mater. Sci. Eng. **21** (2013). https://doi.org/10.1088/0965-0393/21/2/025008
17. Drexler, A., Siegl, W., Ecker, W., Tkadletz, M., Klösch, G., Schnideritsch, H., Mori, G., Svoboda, J., Fischer, F.D.: Cycled hydrogen permeation through Armco iron—a joint experimental and modeling approach. Corros. Sci. **176**, 109017 (2020). https://doi.org/10.1016/j.corsci.2020.109017
18. Siegl, W., Ecker, W., Klarner, J., Kloesch, G., Mori, G., Drexler, A., Winter, G., Schnideritsch, H.: Hydrogen trapping in heat treated and deformed Armco iron. In: NACE - International Corrosion Conference Series, pp. 1–12 (2019)
19. Kholtobina, A.S., Pippan, R., Romaner, L., Scheiber, D., Ecker, W., Razumovskiy, V.I.: Hydrogen Trapping in bcc Iron. Materials (Basel). **13**, 2288 (2020). https://doi.org/10.3390/ma13102288
20. Counts, W.A., Wolverton, C., Gibala, R.: First-principles energetics of hydrogen traps in α-Fe: point defects. Acta Mater. **58**, 4730–4741 (2010). https://doi.org/10.1016/j.actamat.2010.05.010
21. Kiuchi, K., McLellan, R.B.: The solubility and diffusivity of hydrogen in well-annealed and deformed iron. Acta Metall. **31**, 961–984 (1983). https://doi.org/10.1016/0001-6160(83)90192-X
22. Völkl, J., Alefeld, G.: Diffusion of hydrogen in metals. In: Hydrogen in Metals I. Basic Properties, pp. 321–348 (1978)
23. Liu, X., Xie, W., Chen, W., Zhang, H.: Effects of grain boundary and boundary inclination on hydrogen diffusion in α-iron. J. Mater. Res. **26**, 2735–2743 (2011). https://doi.org/10.1557/jmr.2011.262
24. Hagi, H.: Diffusion coefficient of hydrogen in iron without trapping by dislocations and impurities. Mater. Trans. JIM. **35**, 112–117 (1994). https://doi.org/10.2320/matertrans1989.35.112
25. Toribio, J., Kharin, V.: A generalised model of hydrogen diffusion in metals with multiple trap types. Philos. Mag. **95**, 3429–3451 (2015). https://doi.org/10.1080/14786435.2015.1079660
26. Oriani, R.: The diffusion and trapping of hydrogen in steel. Acta Metall. **18**, 147–157 (1970). https://doi.org/10.1016/0001-6160(70)90078-7
27. Iino, M.: Analysis of irreversible hydrogen trapping. Acta Metall. **30**, 377–383 (1982). https://doi.org/10.1016/0001-6160(82)90217-6
28. Drexler, A., Depover, T., Verbeken, K., Ecker, W.: Model-based interpretation of thermal desorption spectra of Fe–C–Ti alloys. J. Alloys Compd. **789**, 647–657 (2019). https://doi.org/10.1016/j.jallcom.2019.03.102
29. Di Stefano, D., Nazarov, R., Hickel, T., Neugebauer, J., Mrovec, M., Elsässer, C.: First-principles investigation of hydrogen interaction with TiC precipitates in α-Fe. Phys. Rev. B. **93**, 184108 (2016). https://doi.org/10.1103/PhysRevB.93.184108
30. He, Y., Li, Y., Chen, C., Yu, H.: Diffusion coefficient of hydrogen interstitial atom in α-Fe, γ-Fe and ϵ-Fe crystals by first-principle calculations. Int. J. Hydrogen Energy. **42**, 27438–27445 (2017). https://doi.org/10.1016/j.ijhydene.2017.08.212
31. Depover, T., Verbeken, K.: Evaluation of the effect of V_4C_3 precipitates on the hydrogen induced mechanical degradation in Fe-C-V alloys. Mater. Sci. Eng. A. **675**, 299–313 (2016). https://doi.org/10.1016/j.msea.2016.08.053
32. Depover, T., Verbeken, K.: Evaluation of the role of Mo_2C in hydrogen induced ductility loss in Q&T Fe-C-Mo alloys. Int. J. Hydrogen Energy. **41**, 14310–14329 (2016). https://doi.org/10.1016/j.ijhydene.2016.05.176

33. Depover, T., Verbeken, K.: The effect of TiC on the hydrogen induced ductility loss and trapping behavior of Fe-C-Ti alloys. Corros. Sci. **112**, 308–326 (2016). https://doi.org/10.1016/j.corsci.2016.07.013
34. Wallaert, E., Depover, T., Arafin, M., Verbeken, K.: Thermal desorption spectroscopy evaluation of the hydrogen-trapping capacity of NbC and NbN precipitates. Metall. Mater. Trans. A. **45**, 2412–2420 (2014). https://doi.org/10.1007/s11661-013-2181-1
35. Drexler, A., Depover, T., Leitner, S., Verbeken, K., Ecker, W.: Microstructural based hydrogen diffusion and trapping models applied to Fe-C-X alloys. J. Alloys Compd. **826**, 154057 (2020). https://doi.org/10.1016/j.jallcom.2020.154057
36. Drexler, A., Bergmann, C., Manke, G., Kokotin, V., Mraczek, K., Pohl, M., Ecker, W.: On the local evaluation of the hydrogen susceptibility of cold-formed and heat treated advanced high strength steel (AHSS) sheets. Mater. Sci. Eng. A. **800** (2021). https://doi.org/10.1016/j.msea.2020.140276
37. Song, E.J., Suh, D.W., Bhadeshia, H.K.D.H.: Theory for hydrogen desorption in ferritic steel. Comput. Mater. Sci. **79**, 36–44 (2013). https://doi.org/10.1016/j.commatsci.2013.06.008
38. Somerday, B.P., Dadfarnia, M., Balch, D.K., Nibur, K.A., Cadden, C.H., Sofronis, P.: Hydrogen-assisted crack propagation in austenitic stainless steel fusion welds. Metall. Mater. Trans. A. **40**, 2350–2362 (2009). https://doi.org/10.1007/s11661-009-9922-1
39. Kumnick, A.J., Johnson, H.H.: Deep trapping states for hydrogen in deformed iron. Acta Metall. **28**, 33–39 (1980). https://doi.org/10.1016/0001-6160(80)90038-3
40. Drexler, A., Ecker, W., Winzer, N., Mraczek, K., Kokotin, V., Manke, G., Bergmann, C.: A step towards numerical evaluation of the local hydrogen susceptibility of punched and cold-formed advanced high strength steel (AHSS) sheets. In: SteelyHydrogen, p. A02 (2018)
41. Jiang, D.E., Carter, E.A.: Diffusion of interstitial hydrogen into and through bcc Fe from first principles. Phys. Rev. B. **70**, 064102 (2004). https://doi.org/10.1103/PhysRevB.70.064102
42. Drexler, A., Bergmann, C., Kokotin, V., Mraczek, K., Manke, G., Leitner, S., Pohl, M., Ecker, W.: Local hydrogen accumulation after cold forming and heat treatment in punched advanced high strength steel sheets. J. Alloys Compd. 158226 (2020). https://doi.org/10.1016/j.jallcom.2020.158226
43. Grabke, H.J., Riecke, E.: Absorption and diffusion AF hydrogen in steels. Mater. Tehnol. **34**, 331–342 (2000)
44. Novak, P., Ritchie, R.O., Sofronis, P., Somerday, B.P., Yuan, R.: A statistical, physical-based, micro-mechanical model of hydrogen-induced intergranular fracture in steel. J. Mech. Phys. Solids. **58**, 206–226 (2009). https://doi.org/10.1016/j.jmps.2009.10.005
45. Hirth, J.P.: Effects of hydrogen on the properties of iron and steel. Metall. Trans. A. **11**, 861–890 (1980). https://doi.org/10.1007/BF02654700
46. Fischer, F.D., Svoboda, J.: Formation of bubbles by hydrogen attack and elastic-plastic deformation of the matrix. Int. J. Plast. **63**, 110–123 (2014). https://doi.org/10.1016/j.ijplas.2013.10.007
47. Vecchi, L., Simillion, H., Montoya, R., Van Laethem, D., Van den Eeckhout, E., Verbeken, K., Terryn, H., Deconinck, J., Van Ingelgem, Y.: Modelling of hydrogen permeation experiments in iron alloys: characterization of the accessible parameters - Part I - The entry side. Electrochem. Acta. **262**, 57–65 (2018). https://doi.org/10.1016/j.electacta.2017.12.172
48. Vecchi, L., Simillion, H., Montoya, R., Van Laethem, D., Van den Eeckhout, E., Verbeken, K., Terryn, H., Deconinck, J., Van Ingelgem, Y.: Modelling of hydrogen permeation experiments in iron alloys: Characterization of the accessible parameters - Part II - The exit side. Electrochem. Acta. **262**, 153–161 (2018). https://doi.org/10.1016/j.electacta.2017.12.173
49. Kirchheim, R.: Solid solutions of hydrogen in complex materials. In: Solid State Physics - Advances in Research and Applications, pp. 203–291 (2004)
50. Svoboda, J., Mori, G., Prethaler, A., Fischer, F.D.: Determination of trapping parameters and the chemical diffusion coefficient from hydrogen permeation experiments. Corros. Sci. **82**, 93–100 (2014). https://doi.org/10.1016/j.corsci.2014.01.002
51. Drexler, A., He, S., Razumovskiy, V., Romaner, L., Ecker, W., Pippan, R.: Verification of the generalized chemical potential for stress driven hydrogen diffusion in nickel. Philos. Mag. Lett. **100**, 513–523 (2020). https://doi.org/10.1080/09500839.2020.1808253

52. Scully, J.R., Young, G.A., Smith, S.W.: Hydrogen embrittlement of aluminum and aluminum-based alloys. In: Gaseous Hydrogen Embrittlement of Materials in Energy Technologiesm, pp. 707–768. Elsevier (2012)
53. Venezuela, J., Liu, Q., Zhang, M., Zhou, Q., Atrens, A.: A review of hydrogen embrittlement of martensitic advanced high-strength steels. Corros. Rev. **34**, 153–186 (2016). https://doi.org/10.1515/corrrev-2016-0006
54. Yamamoto, S.: Delayed fracture. Tetsu-to-Hagane **55**, 326–337 (1969). https://doi.org/10.2355/tetsutohagane1955.55.4_326
55. Ozdirik, B., Suter, T., Hans, U., Depover, T., Verbeken, K., Schmutz, P., Jeurgens, L.P.H., Terryn, H., De Graeve, I.: Study of the hydrogen uptake in deformed steel using the microcapillary cell technique. Corros. Sci. **155**, 55–66 (2019). https://doi.org/10.1016/j.corsci.2019.04.029
56. Polyanskiy, V.A., Belyaev, A.K., Alekseeva, E.L., Polyanskiy, A.M., Tretyakov, D.A., Yakovlev, Y.A.: Phenomenon of skin effect in metals due to hydrogen absorption. Contin. Mech. Thermodyn. **31**, 1961–1975 (2019). https://doi.org/10.1007/s00161-019-00839-2

Chapter 7
Effective Diffusion Coefficient of a Porous Material Applied to the Problem of Hydrogen Damage

Ksenia P. Frolova and Elena N. Vilchevskaya

Abstract The chapter focuses on calculation of the effective diffusion coefficient of a porous material accounting for the volume fraction, shape of pores, and their distribution over orientations in a three-dimensional solid. The existing pores are considered as embedded inhomogeneities possessing a high diffusivity in comparison with a matrix. The segregation effect is taken into account. Maxwell homogenization schemes in terms of diffusivity and resistivity contribution tensors are used. Inhomogeneities are assumed to have a spheroidal shape. The paper considers diverse microstructural patterns, namely, (1) arbitrary orientation distribution of pores, (2) orientational scatter of pores about a preferential orientation, (3) arbitrary orientation distribution of rotational axes of spheroidal pores in one plane. Application of the model to problems related to hydrogen damage is discussed.

Keywords Effective diffusivity · Homogenization problem · Maxwell scheme · Segregation effect

7.1 Introduction

Correct estimation of the value of the diffusion coefficient allows one to predict the amount of hydrogen in metals and alloys. Increasing the amount of hydrogen, in turn, may lead to a premature failure of structural elements of metal parts [1, 2]. Determination of the effective diffusion coefficient of a porous material is of practical interest, since voids and microcracks are known to be traps for hydrogen [3, 4].

K. P. Frolova (✉) · E. N. Vilchevskaya
Institute for Problems in Mechanical Engineering RAS, Bolshoy pr., 61, V.O., 199178 St. Petersburg, Russia
e-mail: kspfrolova@gmail.com

E. N. Vilchevskaya
e-mail: vilchevska@gmail.com

V. A. Polyanskiy and A. K. Belyaev (eds.), *Advances in Hydrogen Embrittlement Study*, Advanced Structured Materials 143, https://doi.org/10.1007/978-3-030-66948-5_7

Diffusion of hydrogen in materials containing microcracks and discontinuities formed during manufacture can be observed in sacrificial coatings that are used to protect steel against corrosion [5, 6]. Among other reasons, permeability of coatings is determined by its microstructure. In [5], it was shown that Zn–Ni coatings can contain through-thickness defects that makes the steel exposed to hydrogen uptake. Correlation between the diffusion coefficient and the microstructural characteristics of Zn–Ni and Cd coatings was investigated in [6]. Thus, it is of quite importance to predict permeability of materials containing pores emerged before hydrogenation.

Numerous papers are devoted to investigation of microstructure of materials affected by hydrogen [1, 7–11]. It is believed that hydrogen diffuses through metals and alloys lattice and accumulates within structural defects such as dislocations, pores, and vacancies leading to initiation and propagation of hydrogen-stimulated microcracks. Typically, defects are occurring at grain boundaries or inclusions due to manufacturing. Thus, in the absence of the inner and outer stresses, hydrogen-induced cracking is observed to take place mainly along grain boundaries [1, 7, 9]. Two types of intergranular cracking can be found in material, namely, void formation along grain boundaries and grain boundary triple junction cracking [1, 7, 10, 11]. Also, hydrogen charging may lead to blistering of the surface [8, 11]. The hydrogen-induced cracks are oriented primarily along the rolling direction of steel [8, 9]. Hydrogen-induced defects may, in turn, increase permeability of a material right during hydrogenation.

The focus is usually set on diffusion of hydrogen along grain boundaries while investigating the effect of presence of high diffusivity paths in metals. To the best of our knowledge, the earliest quantitative model for a diffusion in a heterogeneous material accounting for a single rectilinear grain boundary embedded in a semi-infinite solid of much lower diffusivity was proposed in [12] for the problem of self-diffusion of silver. The same geometry with less restrictive boundary conditions has been considered in [13, 14]. Lamellar and columnar microstructures were modeled in [15], where the obtained expressions were similar to Wiener bounds for thermal or electrical conductivity [16]. The fact that the grain boundaries completely surround grains in a three-dimensionally interconnected network, and are not simply distributed in parallel, has been addressed by numerous diffusion models that embedded grains in a matrix representing grain boundaries [17, 18]. A rule of mixture was used in [19] to calculate the effective diffusion coefficient. A number of approximate schemes accounting for the interactions between inhomogeneities were rewritten in [20] for diffusivity on the base of similarity between governing equations in the diffusivity and conductivity problems. However, a principal difference between two problems is that temperature is a continuous function across the phase boundaries, while concentration is usually not. Therefore, the segregation effect should be taken into account [21]. Hart and Maxwell–Garnett equations were rewritten in the form containing the segregation factor in [22, 23]. The results may be employed for description of materials containing spherical grains. Typical micromechanical models were rewritten for diffusivity in [24] to calculate the effective diffusion coefficient of a polycrystalline material accounting for the isotropic distribution of spheroidal grains over orientation. A similar problem was solved in [25] numerically.

The present paper is concerned with a quantitative characterization of mass transport process in a porous material. Such a microstructural pattern is corresponding to a composite material consisting of grains, grain boundaries, various defects, and isolated pores. Hydrogen is assumed to diffuse along grain boundaries and fill the pores. A composite consisting of grains, grain boundaries, and various defects is represented by a homogenized background matrix with a known effective diffusivity. Pores are considered as embedded inhomogeneities.

The effect of oblate and prolate spheroidal pores, as well as spherical ones, is investigated. An oblate spheroidal pore models a hydrogen-stimulated microcrack occurring between two grains. A prolate spheroidal inhomogeneity reflects the influence of an intragranular microcrack or microcrack propagating along a few grain boundaries. Such a microstructure can be observed in non-saturated materials like sacrificial coatings. Generally speaking, in this case, it is more correct to introduce an ellipsoidal inhomogeneity that is not considered within the frame of the present research. A spherical inhomogeneity models an isolated pore in the host material.

The paper accounts for different orientation distribution of inhomogeneities, namely,

- Arbitrary orientation distribution of pores specific for weakly deformable materials;
- Orientational scatter of pores about a preferential orientation specific for materials prepared by rolling;
- Arbitrary orientation distribution of rotational axes of spheroidal pores in one plane that can be observed for compressed materials.

All the mentioned cases are corresponding to the transversally isotropic porous material. The question arises, how the diffusion coefficient of a matrix material containing grains, grain boundaries, and various defects changes due to the presence of pores.

7.2 Problem Statement

In the present paper, we apply micromechanical approximate schemes to calculate effective diffusivity of a material containing isolated pores assuming that the matrix material (consisting of grains, grain boundaries, and various defects) is homogeneous. Thus, homogenization contains two steps. At the first implicit step, the effective properties of matrix material are obtained by means of experimental data [26] or theoretical framework [24]. At the second explicit step, the effective properties of a porous material are calculated (see Fig. 7.1).

To reflect the effect of pores on the effective properties of a heterogeneous material, we consider inhomogeneities with diffusion coefficient D_1 in a solid of much lower diffusion coefficient D_0. The value $D_1 \to \infty$ providing a zero ratio $\alpha = D_0/D_1$ can be considered for simpleness. However, it is more correct to take into account a non-zero $\alpha \ll 1$ to explain a real phenomenon. We assume that the value of the

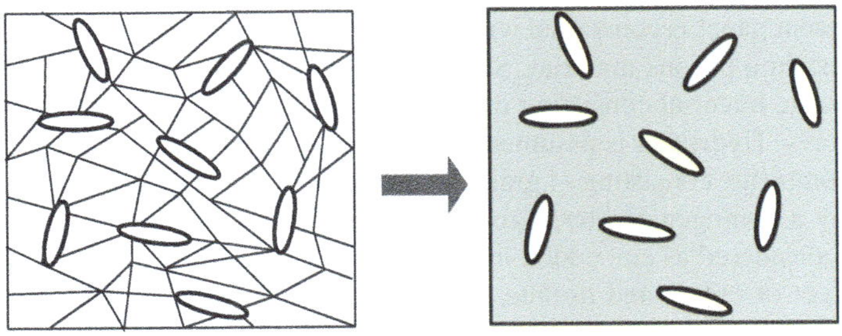

Fig. 7.1 Schematic representation of the model

diffusion coefficient of pores is the same as for unfilled grain boundaries ($D_1 = D_{GB}$). According to [27], the contrast in the diffusion coefficient of the bulk material of grains, D_B, and grain boundaries, D_{GB}, varies from 0.005 to 0.1. Knyazeva et al. [24] considered $\widetilde{\alpha} = D_B/D_{GB} = 0.015$. The contrast in the effective diffusion coefficient of matrix, D_0, and bulk material of grains, D_B, increases with decreasing of the grain size [24, 26]. If the grain diameter is about 10 – 13 μm, the value of $\widetilde{\alpha}_{eff} = D_0/D_B$ is equal to 2.2–4.4 [24, 26]. Then,

$$\alpha = \frac{D_0}{D_1} = \widetilde{\alpha}_{eff}\widetilde{\alpha}$$

takes values in the range 0.03–0.07. Within the frame of the present paper, we assume that $\alpha = 0.05$.

We assume continuity of the normal component of the solute flux across the matrix/pore interface and a constant jump in hydrogen concentration, c, described by the segregation factor, s, as follows:

$$c(x)\,|_{x\to\partial V+} = sc(x)\,|_{x\to\partial V-} \tag{7.1}$$

We are especially interested in cases when hydrogen is partially trapped inside the pores and, therefore, focus on cases $s < 1$. For completeness, we also consider the case when concentration is a continuous function across the phase boundaries and, therefore, there is no segregation effect ($s = 1$). We assume that the segregation effect is independent of the type of pore. Also, segregation related to pore saturation is not discussed here.

The certain preferential orientation distribution of pores is described by means of the probability density function introduced by Sevostianov [28]:

$$\psi_\lambda(\theta) = \frac{1}{2\pi}\left[\left(\lambda^2+1\right)e^{-\lambda\theta} + \lambda e^{-\lambda\pi/2}\right]. \tag{7.2}$$

This probability density function is defined on the upper semisphere of unit radius ($0 \le \theta \le \pi/2$) and subject to the normalization condition. The scatter parameter λ

varies in the range from zero to infinity that corresponds to fully random and strictly parallel orientations of pores, respectively.

To obtain the explicit expressions for the effective properties, we use the formulation of Maxwell's homogenization scheme in terms of diffusivity (resistivity) contribution tensors [24, 29].

7.3 Diffusivity Contribution Tensors

Following [29, 30], we introduce property contribution tensors to express the effect of a given inhomogeneity on the properties of interest. Their sums are proper microstructural parameters that reflect contributions of individual inhomogeneities to the overall effective properties. To explain the process of mass transfer, we introduce the second-rank diffusivity and resistance contribution tensors, $\mathbf{H}^D$ or $\mathbf{H}^{DR}$.

Let us consider a reference volume V of an infinite three-dimensional solid with the isotropic diffusivity tensor $\mathbf{D}_0 = D_0\mathbf{I}$ containing inhomogeneity with the isotropic diffusivity tensor $\mathbf{D}_1 = D_1\mathbf{I}$ occupying domain $V_1 \ll V$. We assume that both inhomogeneity and the surrounding material obey the linear Fick's law connecting the vector of molar flux with the concentration gradient. The extra molar flux due to the inhomogeneity is as follows:

$$\Delta\mathbf{J} = \frac{V_1}{V}\mathbf{H}^D \cdot \mathbf{G}_0, \tag{7.3}$$

where $\mathbf{G}_0$ is a concentration gradient prescribed at the boundary of V. The diffusivity contribution tensor $\mathbf{H}^D$ depends on the shape of the inhomogeneity, diffusivity contrast between inhomogeneity and surrounding material, and the segregation of particles at the interface.

Assuming continuity of the normal component of the solute flux across the matrix/inhomogeneity interface and a segregation effect described by Eq. (7.1), the diffusivity contribution tensor representing the contribution of the spheroidal inhomogeneity to the overall diffusivity of volume V can be written in the form [24]

$$\mathbf{H}^D = D_0\left[B_1\left(\mathbf{I} - \mathbf{nn}\right) + B_2\mathbf{nn}\right], \tag{7.4}$$

where taking $\alpha = D_0/D_1$,

$$B_1 = \frac{1-\alpha}{s\alpha + (1-s\alpha)\,f_0}, \qquad B_2 = \frac{1-\alpha}{1-2\,(1-s\alpha)\,f_0}. \tag{7.5}$$

Shape function $f_0 = f_0\,(\gamma)$ depends on the aspect ratio of the spheroidal inhomogeneity $\gamma = a_3/a$ (a_3 is the rotation axis, $a_1 = a_2 = a$ are two other semiaxes) in the following way:

$$f_0 = \frac{1-g}{2\left(1-\gamma^{-2}\right)}, \tag{7.6}$$

where

$$g = \begin{cases} \frac{1}{\gamma\sqrt{1-\gamma^2}} \arctan \frac{\sqrt{1-\gamma^2}}{\gamma}, \gamma \leq 1 \\ \frac{1}{2\gamma\sqrt{\gamma^2-1}} \ln\left(\frac{\gamma+\sqrt{\gamma^2-1}}{\gamma-\sqrt{\gamma^2-1}}\right), \gamma \geq 1. \end{cases} \tag{7.7}$$

The diffusion resistance contribution tensor $\mathbf{H}^{DR}$ that is dual to $\mathbf{H}^{D}$ can be introduced in a similar way. It appears if, instead of uniform gradient of concentration, a uniform molar flux is remotely applied. Tensors $\mathbf{H}^{DR}$ and $\mathbf{H}^{D}$ interrelate as follows [24]:

$$\mathbf{H}^{DR} = -\frac{1}{D_0^2}\mathbf{H}^{D}. \tag{7.8}$$

According to [29], there are no any explicit evidence regarding preference of one of tensor $\mathbf{H}^{DR}$ or $\mathbf{H}^{D}$. So, it is necessary to check which one of these tensors yields better agreement with experimental data.

Note that according to Eqs. (7.4) and (7.5), the segregation factor s does not play any role and simplifies the model in the case when $\alpha \to 0$ due to $D_1 \to \infty$.

7.4 Homogenization Problem

An exact solution of the homogenization problem for a composite material with multiple inhomogeneities can be obtained only numerically. Moreover, specific materials with a known microstructure should be considered. All analytical approximation methods involve uncertainties and/or inconsistencies. A detailed review of history of various methods can be found in [31], whereas the current state of knowledge of the problem is described in [29].

We use the Maxwell homogenization scheme, since it appears to be one of the best schemes, in terms of its applicability to cases of multiphase composites and accuracy. In his original work, Maxwell addressed the problem of effective electrical conductivity of a material containing multiple spherical inhomogeneities [32]. The accuracy of Maxwell's original formula was checked by means of a periodic array of spheres and found to be accurate up to volume fractions about 40% [33]. We use the formulation of Maxwell scheme in terms of property contribution tensors proposed in [28, 34].

According to Maxwell's idea, it is necessary to evaluate far-field perturbations due to inhomogeneities in two different ways and equate the results. The first way is to evaluate this field as the one generated by a homogenized region Ω possessing the (yet unknown) effective properties. This field can be expressed in terms of the property contribution tensor of the domain Ω. The second way is based on consideration the sum of far fields generated by all the individual inhomogeneities within Ω (treated as non-interacting ones). Equating the results yields the effective diffusion properties in the form [24]

$$\left(\mathbf{D}^{eff}\right)^{-1} = \frac{1}{D_0}\mathbf{I} + \left\{\left[\frac{1}{V_\Omega}\sum_k V_k \mathbf{H}_k^{DR}\right]^{-1} - \mathbf{Q}_\Omega^D\right\}^{-1} \tag{7.9}$$

in terms of the resistivity and

$$\mathbf{D}^{eff} = D_0\mathbf{I} + \left\{\left[\frac{1}{V_\Omega}\sum_k V_k \mathbf{H}_k^{D}\right]^{-1} - \mathbf{P}_\Omega^D\right\}^{-1} \tag{7.10}$$

in terms of diffusivity. Here,

$$\mathbf{Q}_\Omega^D = D_0 s_\Omega\left[\left(1 - f_0\left(\gamma_\Omega^D\right)\right)(\mathbf{I} - \mathbf{mm}) + 2f_0\left(\gamma_\Omega^D\right)\mathbf{mm}\right], \tag{7.11}$$

$$\mathbf{P}_\Omega^D = \frac{1}{D_0 s_\Omega}\left[f_0\left(\gamma_\Omega^D\right)(\mathbf{I} - \mathbf{mm}) + (1 - 2f_0)\left(\gamma_\Omega^D\right)\mathbf{mm}\right] \tag{7.12}$$

are the second-rank Hill's tensors that reflect the shape of spheroidal domain Ω and take into account interactions between the inhomogeneities. Vector **m** is supposed to be a unit vector along the axis of symmetry of the domain Ω. Note that **n** is corresponding to the axis of symmetry of individual spheroidal inhomogeneity and, therefore, **m** may not coincidence with **n**.

The segregation factor s_Ω describes a constant jump in the particles concentration at the interface between the domain Ω and matrix. This segregation factor can not be estimated experimentally in contrast to the segregation factor for the matrix/inhomogeneity interface. In [35, 36], these parameters were supposed to be equal. Results obtained at $s_\Omega = s$ and $s_\Omega = 1$ were compared with the experimental data for a polycrystalline material in [24]. It was shown that $s_\Omega = 1$ does not provide correct values of the effective diffusion coefficient at large values of the volume fraction of inhomogeneities. At the same time, value of s_Ω does not play a role at small values of the volume fraction of inhomogeneities.

Recommendations regarding the choice of the shape of the domain Ω can be found in [28]. This shape has to obey the following requirements: (1) it should be ellipsoidal, (2) it should properly reflect the shapes of individual inhomogeneities, their orientations and properties. Within the frame of the present paper, we consider materials with transversely isotropic microstructure. In this case, the domain Ω is a spheroid. In the specific case of random orientation distribution of inhomogeneities, the shape of the domain is spherical ($\gamma_\Omega = 1$). In the case of preferentially oriented inhomogeneities, following [28], we introduce the shape of the domain Ω as

$$\gamma_{\Omega}^{D} = \frac{\sum_{k} V_k P_{11}^{D}}{\sum_{k} V_k P_{33}^{D}}, \tag{7.13}$$

where P_{11}, P_{33} are components of the second-rank Hill's tensor

$$\mathbf{P}^{D} = \frac{1}{D_0 s}\left[f_0(\gamma)(\mathbf{I} - \mathbf{nn}) + (1 - 2f_0(\gamma))\,\mathbf{nn}\right] \tag{7.14}$$

for individual spheroidal inhomogeneity.

The effective properties of a considered material with pores of identical shapes and size and diverse distribution over their orientations can be expressed in terms of their volume fraction $\phi = \sum_k V_k / V$ [29]. Summation in Eqs. (7.9), (7.10), and (7.13) can be replaced by multiplication of the fraction volume by averaged tensors, so

$$\frac{1}{V}\sum_{k} V_k \mathbf{H}^{D} = \phi\left\langle \mathbf{H}^{D}\right\rangle, \qquad \frac{1}{V}\sum_{k} V_k \mathbf{H}^{DR} = \phi\left\langle \mathbf{H}^{DR}\right\rangle,$$

$$\frac{1}{V}\sum_{k} V_k \mathbf{P}^{D} = \phi\left\langle \mathbf{P}^{D}\right\rangle.$$

The tensor averaging is equivalent to averaging of dyads **nn**, since these tensors are determined by means of **nn** and $\mathbf{I} - \mathbf{nn}$.

We first consider orientation distribution of **n** about a preferential orientation **m** accompanied by a random scatter. This orientation distribution of pores is described by Eq. (7.2). The averaging of the dyad **nn** can be obtained by integration over the upper semisphere of unit radius $\tilde{\Omega}_{1/2}$, so

$$\langle \mathbf{nn} \rangle = \int_{\tilde{\Omega}_{1/2}} (\mathbf{nn})_k \psi_\lambda d\tilde{\Omega}_{1/2}.$$

Then,

$$\langle \mathbf{nn} \rangle = g_1 \boldsymbol{\theta} + g_2 \mathbf{mm},$$

where $\boldsymbol{\theta} = \mathbf{I} - \mathbf{mm}$ and

$$g_1 = \frac{18 - e^{-\frac{\pi\lambda}{2}}\lambda\left(3 + \lambda^2\right)}{6\left(9 + \lambda^2\right)}, \qquad g_2 = \frac{\left(3 + e^{-\frac{\pi\lambda}{2}}\lambda\right)\left(3 + \lambda^2\right)}{3\left(9 + \lambda^2\right)}.$$

Equations (7.9), (7.10) reduce to

$$\frac{\mathbf{D}^{eff}}{D_0} = \frac{1}{1 - \left[\frac{1}{\phi(B_1(1-g_1)+B_2 g_1)} + s_\Omega\left(1 - f_0\left(\gamma_\Omega^D\right)\right)\right]^{-1}}\boldsymbol{\theta} + \\ + \frac{1}{1 - \left[\frac{1}{\phi(B_1(1-g_2)+B_2 g_2)} + 2s_\Omega f_0\left(\gamma_\Omega^D\right)\right]^{-1}}\mathbf{mm}, \tag{7.15}$$

$$\frac{\mathbf{D}^{eff}}{D_0} = \left(1 + \left[\frac{1}{\phi\left(B_1\left(1 - g_1\right) + B_2 g_1\right)} - \frac{f_0\left(\gamma_\Omega^D\right)}{s_\Omega}\right]^{-1}\right)\boldsymbol{\theta} + \\ + \left(1 + \left[\frac{1}{\phi\left(B_1\left(1 - g_2\right) + B_2 g_2\right)} - \frac{\left(1 - 2 f_0\left(\gamma_\Omega^D\right)\right)}{s_\Omega}\right]^{-1}\right)\mathbf{mm}. \tag{7.16}$$

In the case of arbitrary orientation distribution of inhomogeneities $\lambda = 0$ and, therefore,

$$\langle \mathbf{nn} \rangle = \frac{1}{3}\mathbf{I}$$

and Eqs. (7.9), (7.10) take forms

$$\frac{D^{eff}}{D_0} = \frac{3 + 2\phi\eta s_\Omega}{3 + \phi\eta\left(2s_\Omega - 3\right)}, \tag{7.17}$$

$$\frac{D^{eff}}{D_0} = \frac{3s_\Omega + \phi\eta\left(3s_\Omega - 1\right)}{3s_\Omega - \phi\eta}, \tag{7.18}$$

where $\eta = \eta\left(\gamma, \alpha, s\right) = 2B_1/3 + B_2/3$.

Now, turn to the case of random orientation distribution of unit vectors $\mathbf{n}$ in the plane normal to unit vector $\mathbf{m}$. The averaging of the dyad $\mathbf{nn}$ can be obtained by integration over a circle of unit radius l_1 normal to $\mathbf{m}$, so

$$\langle \mathbf{nn} \rangle = \frac{1}{2\pi} \int\limits_{l_1} (\mathbf{nn})_k dl_1.$$

Therefore,

$$\langle \mathbf{nn} \rangle = \frac{1}{2}\boldsymbol{\theta}$$

and Eqs. (7.9), (7.10) reduce to

$$\frac{\mathbf{D}^{eff}}{D_0} = \frac{1}{1 - \left[\frac{2}{\phi(B_1+B_2)} + s_\Omega\left(1 - f_0\left(\gamma_\Omega^D\right)\right)\right]^{-1}}\boldsymbol{\theta} + \\ + \frac{1}{1 - \left[\frac{1}{\phi B_1} + 2s_\Omega f_0\left(\gamma_\Omega^D\right)\right]^{-1}}\mathbf{mm}, \tag{7.19}$$

$$\frac{\mathbf{D}^{eff}}{D_0} = \left(1 + \left[\frac{2}{\phi\left(B_1 + B_2\right)} - \frac{f_0\left(\gamma_\Omega^D\right)}{s_\Omega}\right]^{-1}\right)\boldsymbol{\theta} + \\ + \left(1 + \left[\frac{1}{\phi B_1} - \frac{1 - 2f_0\left(\gamma_\Omega^D\right)}{s_\Omega}\right]^{-1}\right)\mathbf{mm}. \tag{7.20}$$

According to [34], Maxwell's scheme formulated in terms of diffusivity and resistivity contribution tensors should lead to the same result. Any contradictions may cause incorrect accounting for interactions between inhomogeneities within the scheme. Comparing the results obtained within the present section in terms of diffusivity and resistivity contribution tensors, one can see that the equality of two solutions imposes restrictions on the segregation factor s_Ω for the interface between the domain Ω and matrix. Maxwell's scheme formulated in terms of diffusivity and resistivity contribution tensors for the considered cases of the orientation distribution of inhomogeneities leads to the same result only at certain values of the segregation factor s_Ω. In particular, effective properties obtained within Maxwell's scheme formulated in terms of diffusivity and resistivity contribution tensors are the same in all the mentioned cases of the orientation pattern when the segregation factor $s_\Omega = 1$ that means that there is no segregation effect at the interface domain Ω/matrix. In the case of random distribution of the inhomogeneities, the segregation factor s_Ω can also be equal to 0.5. In the cases of preferentially oriented inhomogeneities, accounting for the segregation effect leads to incorrect results.

7.5 Effective Diffusion Coefficient of a Material with Spheroidal Pores

The effective diffusion coefficient of a material containing pores depends on the shape of the inhomogeneities, its volume fraction and distribution over orientations, as well as on the segregation effect. We now investigate these dependences for a heterogeneous material described in Sect. 7.2. Thus, we consider $\alpha = D_0/D_1 = 0.05$ and $s \leq 1$.

We start with modeling of an isotropic distribution of spheroidal inhomogeneities. The accuracy of Maxwell's scheme was found to be determined by volume fraction of

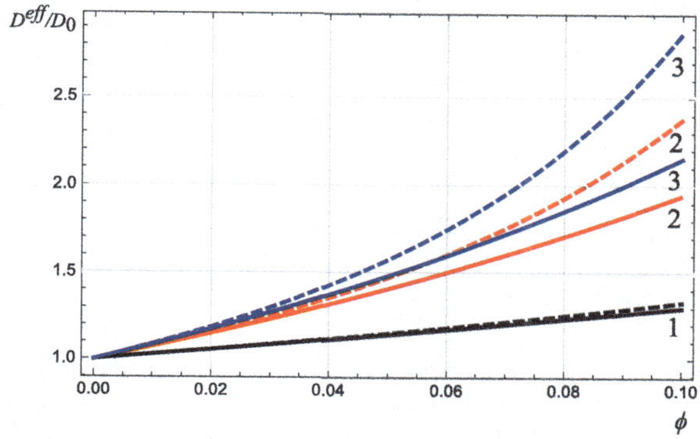

Fig. 7.2 Dependences of the effective diffusion coefficient on the volume fraction of (1) spherical pores, (2) oblate spheroidal pores with $\gamma = 0.1$, (3) prolate spheroidal pores with $\gamma = 10$ at $s_\Omega = 1$ (solid line) and $s_\Omega = 0.5$ (dashed line); case of random orientation distribution of pores

pores. Such a restriction was discussed in our paper [37] for the case of material containing oblate spheroidal pores. Figure 7.2 illustrates dependences of the normalized effective diffusion coefficient, D^{eff}/D_0, on the volume fraction of pores, ϕ, at two different values of the segregation factor for the interface between the inhomogeneity with effective properties occupying domain Ω and matrix ($s_\Omega = 1$ and $s_\Omega = 0.5$). Segregation at the interface between an isolated inhomogeneity and matrix is determined by a fixed value $s = 0.5$. Accounting for the segregation effect at the interface domain Ω/matrix ($s_\Omega \neq 1$) increases the effective diffusion coefficient. Additional efforts are needed to check whether accounting for segregation effect at the interface domain Ω/matrix in the model yields better agreement with the experimental data. Hereafter, we assume that $s_\Omega = 1$.

Figure 7.3 illustrates dependences of the normalized effective diffusion coefficient on the segregation factor at the interface inhomogeneity/matrix at a constant volume fraction $\phi = 0.1$. According to the results, decreasing the segregation factor s increases the effective diffusion coefficient and vice versa. The ratio D^{eff}/D_0 is minimal when there is no segregation effect ($s = 1$). Thus, segregation of particles inside pores produces effect similar to increasing diffusivity of an inhomogeneity. At the same time, it is seen that the segregation of particles inside spherical pores does not change the effective diffusion coefficient significantly. Thus, segregation inside pores increases the effect of shape of the inhomogeneities on the overall diffusion coefficient. Figure 7.4 demonstrates that accounting for segregation of particles inside non-spherical pores at high volume fractions of pores is of great importance. Figure 7.5 illustrates the effect of shape on the effective diffusion coefficient at $\phi = 0.1$ and $s = 0.5$.

Consider now material containing spheroidal pores that have mild tendency to be parallel to each other. Dependences of the effective diffusion coefficients on the scatter parameter are shown in Fig. 7.6. Increasing of the scatter parameter λ, i.e., decreasing of randomness of the orientation distribution, leads to increasing of the effective diffusion coefficient within the plane of isotropy of a material containing oblate spheroidal pores with $\gamma = 0.1$ and to decreasing of its effective diffusion coefficient along the axis of symmetry. In the case of prolate spheroidal pores with

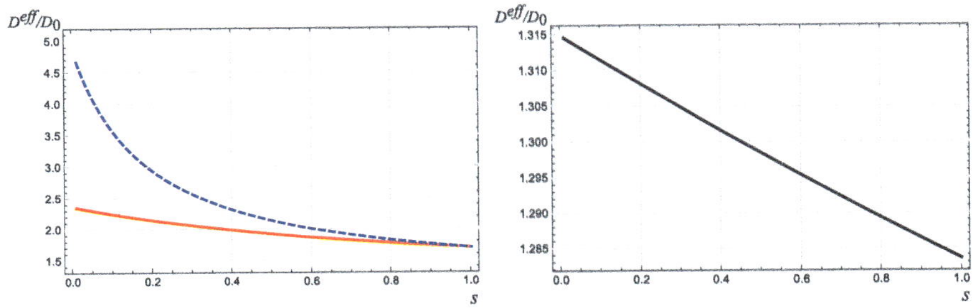

Fig. 7.3 Dependences of the effective diffusion coefficient of a material containing non-spherical pores (left-hand side), namely, oblate spheroidal pores with $\gamma = 0.1$ (solid line) and prolate spheroidal pores with $\gamma = 10$ (dashed line), and spherical pores (right-hand side) on the segregation factor s; case of random orientation distribution of pores

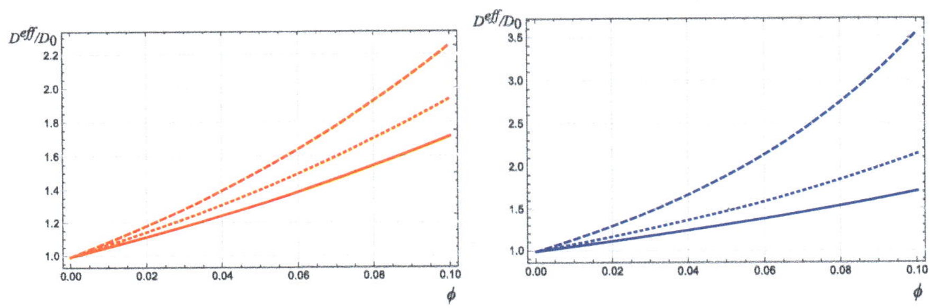

Fig. 7.4 Dependences of the effective diffusion coefficient on the volume fraction of oblate pores with $\gamma = 0.1$ (left-hand side) and prolate pores with $\gamma = 10$ (right-hand side) at $s = 0.1$ (dashed line), $s = 0.5$ (dotted line) and $s = 1$ (solid line); case of random orientation distribution of pores

Fig. 7.5 Dependence of the effective diffusion coefficient on the aspect ratio of pores in the case of random orientation distribution of inhomogeneities

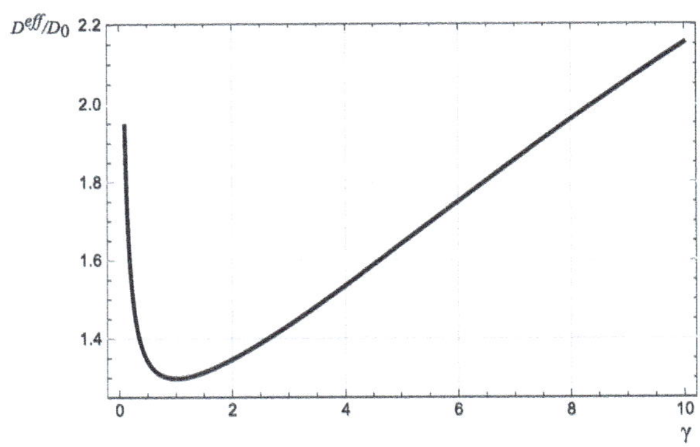

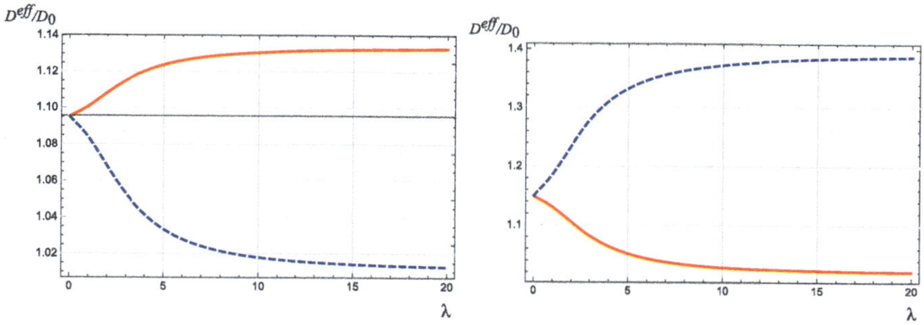

Fig. 7.6 Dependences of the effective diffusion coefficients D_{11}^{eff}/D_0 (solid line) and D_{33}^{eff}/D_0 (dashed line) of a material with oblate spheroidal pores with $\gamma = 0.1$ (left-hand side) and material with prolate spheroidal pores with $\gamma = 10$ (right-hand side) on lambda; case of certain preferential orientation accompanied by random scatter

$\gamma = 10$, it is the other way round. We take $\phi = 0.01$ and $s = 0.5$. Note that the qualitative result will be similar if we use other values of the aspect ratios, volume fraction, and segregation factor.

The oblate spheroidal pores were found to change diffusion within the plane of isotropy and to have no significant influence on the value of the diffusion coefficient along the axis of symmetry. The situation was found to be opposite in the case of prolate spheroids. Figure 7.7 illustrates dependences of the normalized effective diffusion coefficient on the segregation factor at the interface inhomogeneity/matrix at a constant volume fraction $\phi = 0.1$ and scatter parameter $\lambda = 20$. For the sake of brevity, we do not provide dependences D_{33}^{eff}/D_0 on s corresponding to a material with oblate pores and D_{11}^{eff}/D_0 on s corresponding to a material with prolate pores, since these effective diffusion coefficients are weakly dependent on the segregation effect in the mentioned cases due to the weak effect of shape. As in the case of random distribution, segregation inside pores increases effect of shape on the overall diffusion coefficient. Figure 7.8 illustrates the effect of shape on the effective diffusion coefficient at $\phi = 0.1$, $s = 0.5$ and $\lambda = 20$.

Finally, turn to the problem of random orientations of unit vectors **n** in one plane. The oblate spheroidal pores were found to change diffusion within the plane of isotropy and along the axis of symmetry, whereas prolate spheroidal pores were found to influence only diffusivity within the plane of isotropy. Again, for the sake of brevity, we do not provide dependence D_{33}^{eff}/D_0 on s corresponding to a material with prolate pores, since this effective diffusion coefficient weakly depends on the segregation effect in the mentioned case due to the weak effect of shape. Figure 7.9 illustrates dependences of the normalized effective diffusion coefficient on the segregation factor at the interface inhomogeneity/matrix at a constant volume fraction $\phi = 0.1$. It is seen that segregation inside pores increases the effect of shape on the overall diffusion coefficient. Figure 7.10 illustrates the effect of shape on the effective diffusion coefficient.

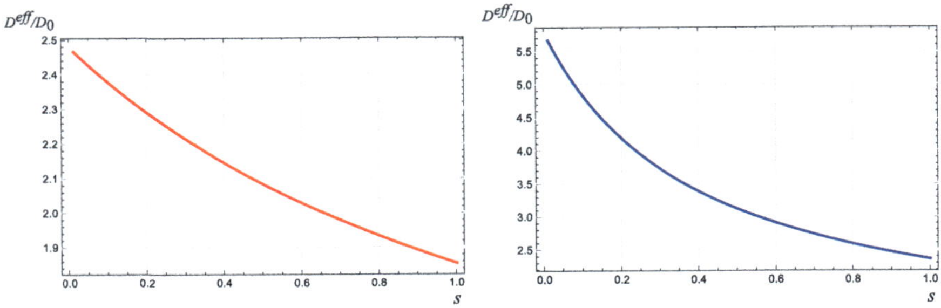

Fig. 7.7 Dependences of the effective diffusion coefficients D_{11}^{eff}/D_0 of a material with oblate spheroidal pores with $\gamma = 0.1$ (left-hand side) and D_{33}^{eff}/D_0 of a material with prolate spheroidal pores with $\gamma = 10$ (right-hand side); case of certain preferential orientation accompanied by a random scatter ($\lambda = 20$)

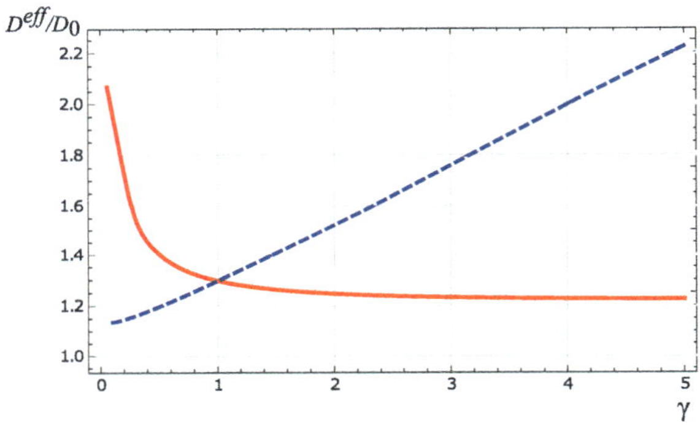

Fig. 7.8 Dependences of the effective diffusion coefficients D_{11}^{eff}/D_0 (solid line) and D_{33}^{eff}/D_0 (dashed line) on the aspect ratio of pores in the case of their certain preferential orientation accompanied by a random scatter ($\lambda = 20$)

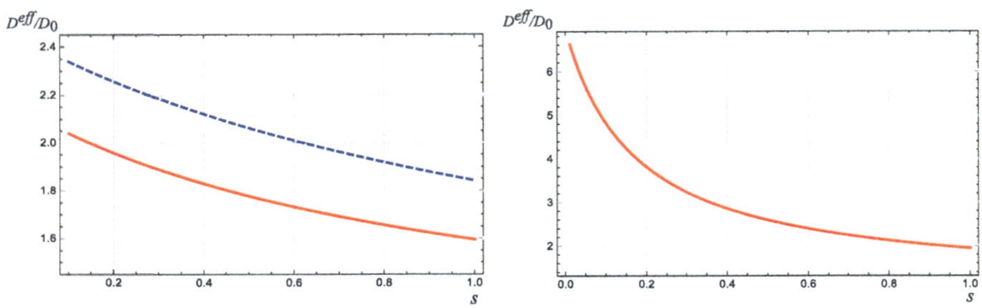

Fig. 7.9 Dependences of the effective diffusion coefficients D_{11}^{eff}/D_0 (solid line) and D_{33}^{eff}/D_0 (dashed line) of a material containing oblate spheroidal pores with $\gamma = 0.1$ (left-hand side) and prolate spheroidal pores with $\gamma = 10$ (right-hand side) on the segregation factor s; case of random orientations of unit vectors **n** in one plane

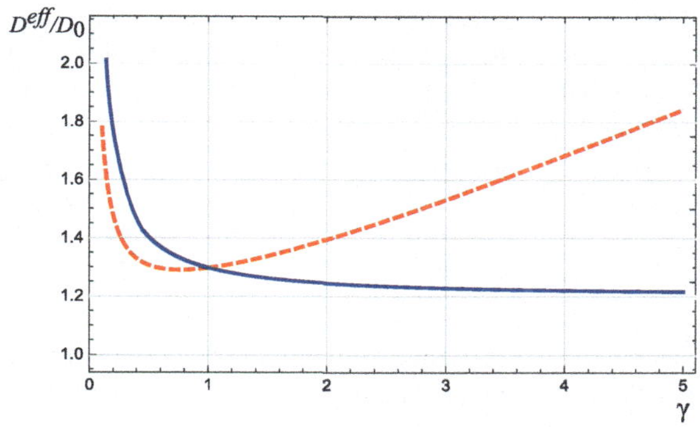

Fig. 7.10 Dependences of the effective diffusion coefficients D_{11}^{eff}/D_0 (dashed line) and D_{33}^{eff}/D_0 (solid line) on the aspect ratio of pores in the case of random orientations of unit vectors **n** in one plane

Thus, it was shown that the presence of multiple pores of diverse shape and distribution over orientation can lead to a significant increase in the diffusion coefficient. The effect of shape was found to be strongly coupled with the distribution of inhomogeneities over orientation. In particular, diffusion coefficients in diverse directions can be changed in various ways with respect to the shape of a non-spherical pore. The segregation of hydrogen at the interface matrix/pore was found to have a significant effect on the effective diffusivity, especially in the case of preferentially oriented prolate pores.

7.6 Description of the Experimental Data by Means of the Proposed Model

To illustrate the applicability of the results of modeling to problems associated with hydrogen damage, we describe results obtained in [6]. Authors of [6] analyzed the permeability of Zn–Ni coating material to hydrogen in correlation with its microstructural characteristics. In particular, they indicated that Zn–Ni coatings typically exhibit defects, such as intergranular cracks and through-thickness pores. Changing of a substrate geometry during the permeability test realized by the twin cell method led to changing of the microcracks distribution in the coatings. As a result, microcracks were found to be randomly distributed. The microstructure of coatings with and without defects was shown in Scanning Electron Microscope (SEM) images. Initiated defects may make coatings susceptible to environmental hydrogen embrittlement. Authors compared the hydrogen diffusion coefficients corresponding to uniform coating and coating with defects. The ratio of the effective diffusion coefficient to the diffusion coefficient of a host material is given in Table 7.1.

Unfortunately, there are no data on the exact shape and volume fraction of the defects. On the base of the SEM images shown in [6], we could say that microcracks appearing in Zn–Ni coating can be considered as prolate spheroidal pores that have certain preferential orientation ($\lambda = 20$). Taking diverse realistic values of γ and

Table 7.1 Correlation of the effective diffusion coefficient D^{eff}/D_0 with its microstructural characteristics, namely, shape (by means of γ), porosity (ϕ), and distribution over orientations

Experimental D^{eff}/D_0	γ (fixed)	ϕ (range of values)	Distribution over orientations	Calculated D^{eff}/D_0
14.8 (Zn-Ni)	20	0.10–0.50	Certain preferential	4.70–19.75
	100	0.10–0.50	orientation ($\lambda = 20$)	3.99 – 16.43

varying the value of the volume fraction, we can obtain a proper range of the effective diffusion coefficient. The results are shown in Table 7.1. We use $s = 0.5$, $s_\Omega = 1$.

The suggested homogenization model leads to results that are in a good agreement with experimental data and, therefore, may be applied to estimate the effective diffusion coefficients of hydrogen in a porous material in the annex to the problem of hydrogen damage.

7.7 Conclusions

The paper addresses the effective diffusion properties of a porous material. The microstructure may comprise a mixture of inhomogeneities of diverse shapes and orientations. We assume that oblate pores can model hydrogen-induced intergranular microcracks, whereas prolate pores can model microcracks originating during processing and affecting further hydrogenation. The segregation of hydrogen inside pores takes place.

Maxwell scheme formulated in terms of property contribution tensors was applied to solve the homogenization problem. It was shown that the presence of pores can lead to a significant increase in the diffusion coefficient. Effect of shape and segregation effect was observed. The homogenization problem was solved at diverse distribution of inhomogeneities over orientations. The effect of shape was found to be coupled with the orientation distribution.

Acknowledgements This study was supported by the Russian Foundation for Basic Research (projects No. 20-08-01100, 18-08-00201).

References

1. Koyama, M., Akiyama, E., Tsuzak, K.: Hydrogen-induced delayed fracture of a Fe–22Mn–0.6 C steel pre-strained at different strain rates. Scripta Mater. **66**(11), 947–950 (2012)
2. So, K.H., Kim, J.S., Chun, Y.S., Park, K.T., Lee, Y.K., Lee, C.S.: Hydrogen delayed fracture properties and internal hydrogen behavior of a Fe–18Mn–1.5 Al–0.6 C TWIP steel. ISIJ Int. **49**(12), 1952–1959 (2009)

3. Nair, S.V., Jensen, R.R., Tien, J.K.: Kinetic enrichment of hydrogen at interfaces and voids by dislocation sweep-in of hydrogen. Metall. Trans. A **14**(2), 385–393 (1983)
4. Pressouyre, G.M.: A classification of hydrogen traps in steel. Metall. Trans. A **10**(10), 1571–1573 (1979)
5. Figueroa, D., Robinson, M.J.: The effects of sacrificial coatings on hydrogen embrittlement and re-embrittlement of ultra high strength steels. Corros. Sci. **50**(4), 1066–1079 (2008)
6. Sriraman, K.R., Brahimi, S., Szpunar, J.A., Yue, S.: Hydrogen embrittlement of Zn-, Zn-Ni-, and Cd-coated high strength steel. J. Appl. Electrochem. **43**(4), 441–451 (2013)
7. Kuhr, B., Farkas, D., Robertson, I.M.: Atomistic studies of hydrogen effects on grain boundary structure and deformation response in FCC Ni. Comput. Mater. Sci. **122**, 92–101 (2016)
8. Merson, E.D., Myagkikh, P.N., Klevtsov, G.V., Merson, D.L., Vinogradov, A.: Effect of fracture mode on acoustic emission behavior in the hydrogen embrittled low-alloy steel. Eng. Fract. Mech. **210**, 342–357 (2019)
9. Shen, C.H., Shewmon, P.G.: A mechanism for hydrogen-induced intergranular stress corrosion cracking in alloy 600. Metall. Trans. A **21**(5), 1261–1271 (1990)
10. Sun, B., Krieger, W., Rohwerder, M., Ponge, D., Raabe, D.: Dependence of hydrogen embrittlement mechanisms on microstructure-driven hydrogen distribution in medium Mn steels. Acta Mater. **183**, 313–328 (2020)
11. Wasim, M., Djukic, M.B.: Hydrogen embrittlement of low carbon structural steel at macro-, micro-and nano-levels. Int. J. Hydrogen Energy **45**(3), 2145–2156 (2020)
12. Fisher, J.C.: Calculation of diffusion penetration curves for surface and grain boundary diffusion. J. Appl. Phys. **22**(11), 74–77 (1951)
13. Le Claire, A.D.: The analysis of grain boundary diffusion measurements. Br. J. Appl. Phys. **14**(6), 351 (1963)
14. Suzuoka, T.: Exact solutions of two ideal cases in grain boundary diffusion problem and the application to sectioning method. J. Phys. Soc. Jpn **19**(6), 839–851 (1964)
15. Gilmer, G.H., Farrell, H.H.: Grain-boundary diffusion in thin films: I. The isolated grain boundary. J. Appl. Phys. **47**(9), 3792–3798 (1976)
16. Wiener, O.: Abhandl. math.-phys. Kl. Königl. Sachsischen Gesell **32**, 509 (1912)
17. Chen, Y., Schuh, C.A.: Geometric considerations for diffusion in polycrystalline solids. J. Appl. Phys. **101**(6), 063524 (2007)
18. Preis, W., Sitte, W.: Surface exchange reactions and fast grain boundary diffusion in polycrystalline materials: Application of a spherical grain model. J. Phys. Chem. Solids **66**(10), 1820–1827 (2005)
19. Hart, E.W.: Thermal conductivity. Acta Metall. **5**, 597–605 (1957)
20. Barrer, R.M.: Diffusion and permeation in heterogeneous media. Diff. Poly. **165–217** (1968)
21. Kaur, I., Mishin, Y., Gust, W.: Fundamentals of Grain and Interphase Boundary Diffusion. Wiley (1995)
22. Belova, I.V., Murch, G.E.: Diffusion in nanocrystalline materials. J. Phys. Chem. Solids **64**(5), 873–878 (2003)
23. Belova, I.V., Murch, G.E.: The effective diffusivity in polycrystalline material in the presence of interphase boundaries. Philos. Mag. **84**(1), 17–28 (2004)
24. Knyazeva, A.G., Grabovetskaya, G.P., Mishin, I.P., Sevostianov, I.: On the micromechanical modelling of the effective diffusion coefficient of a polycrystalline material. Philos. Mag. **95**(19), 2046–2066 (2015)
25. Legrand, E., Bouhattate, J., Feaugas, X., Touzain, S., Garmestani, H., Khaleel, M., Li, D.S.: Numerical analysis of the influence of scale effects and microstructure on hydrogen diffusion in polycrystalline aggregates. Comput. Mater. Sci. **71**, 1–9 (2013)
26. Oudriss, A., Creus, J., Bouhattate, J., Conforto, E., Berziou, C., Savall, C., Feaugas, X.: Grain size and grain-boundary effects on diffusion and trapping of hydrogen in pure nickel. Acta Mater. **60**(19), 6814–6828 (2012)
27. Brass, A.M., Chanfreau, A.: Accelerated diffusion of hydrogen along grain boundaries in nickel. Acta Mater. **44**(9), 3823–3831 (1996)

28. Sevostianov, I.: On the shape of effective inclusion in the Maxwell homogenization scheme for anisotropic elastic composites. Mech. Mater. **75**, 45–59 (2014)
29. Kachanov, M., Sevostianov, I.: Micromechanics of materials, with applications. In: Solid Mechanics and Its Applications, vol. 249. Springer, Cham (2018)
30. Kachanov, M., Sevostianov, I.: On quantitative characterization of microstructures and effective properties. Int. J. Solids Struct. **42**(2), 309–336 (2005)
31. Markov, K.Z.: Elementary micromechanics of heterogeneous media. In: Markov, K., Preziosi, L. (eds.) Heterogeneous Media. Modeling and Simulation in Science, Engineering and Technology, pp. 1–162. Birkhäuser, Boston, MA (2000)
32. Maxwell, J.C.: A Treatise on Electricity and Magnetism. Clarendon Press, Oxford (1873)
33. Rayleigh, L.: LVI. On the influence of obstacles arranged in rectangular order upon the properties of a medium. Philos. Mag. J. Sci. The London, Edinburgh, and Dublin **34**(211), 481–502 (1892)
34. Sevostianov, I., Giraud, A.: Generalization of Maxwell homogenization scheme for elastic material containing inhomogeneities of diverse shape. Int. J. Eng. Sci. **64**, 23–36 (2013)
35. Belova, I.V., Murch, G.E.: Calculation of the effective conductivity and diffusivity in composite solid electrolytes. J. Phys. Chem. Solids **66**(5), 722–728 (2005)
36. Kalnin, J.R., Kotomin, E.A., Maier, J.: Calculations of the effective diffusion coefficient for inhomogeneous media. J. Phys. Chem. Solids **63**(3), 449–456 (2002)
37. Frolova, K.P.: Determination of the effective Young's modulus of medium with microstructure typical for hydrogen degradation. J. Phys. Math. St. Petersburg State Polytechnic University **13**(2), 160 (2020)

Chapter 8
Determination of the Activation Energy of Hydrogen from Their Compounds with Titanium

Polina M. Grigoreva, Yuriy A. Yakovlev, Anatoly M. Polyanskiy, Victor P. Loginov, Andrey V. Polyanskiy, and Vladimir A. Polyanskiy

Abstract In this work, we carry out experiments on the extraction of hydrogen from amorphous titanium hydrides by Vacuum Hot Extraction method. Based on the obtained experimental data, we construct thermal desorption spectra with a stepwise change in the extraction temperature. Using the Kissinger model for decomposition of a solid as for a chemical reaction of the first order, we construct the Arrhenius dependences to calculate the activation energy for the decomposition of titanium hydrides. For the same temperature ranges, the hydrogen binding energies determined for each of the samples differ several times, and the activation energy decreases with increasing temperature. At the same time, for a temperature range of 350–370 °C, the activation energy tends to infinity. This effect commonly is explained by the influence of traps corresponding to the chemical composition and microstructure of the material. We question this explanation, since we study various samples of the same chemical compound, and the resulting activation energies are in the range that does not correspond to the value of the chemical bond energies.

Keywords Hydrogen diffusion · Thermo-desorption Spectra (TDS) · Activation energy · Hydrogen embrittlement · Hydrogen storage

8.1 Introduction

The hydrogen influence on metal properties has been a well-known and studied problem in science since the end of the nineteenth century cf. [1, 2], [?]. The absorption of hydrogen from the external environment leads to the degradation of the mechan-

---

P. M. Grigoreva (✉) · Y. A. Yakovlev · V. A. Polyanskiy
Institute for Problems in Mechanical Engineering RAS, Bolshoy pr., 61, V.O., St. Petersburg 199178, Russia
e-mail: polina.grigoreva239@gmail.com

A. M. Polyanskiy · V. P. Loginov · A. V. Polyanskiy
RDC Electronic and Beam Technologyes, Ltd., Politekhnicheskaya, 29, St. Petersburg 194064, Russia

V. A. Polyanskiy and A. K. Belyaev (eds.), *Advances in Hydrogen Embrittlement Study*, Advanced Structured Materials 143, https://doi.org/10.1007/978-3-030-66948-5_8

ical properties of metal structures, which is known as the problem of hydrogen embrittlement. To control this process, understanding the processes of absorption of hydrogen from the external environment, the diffusion of materials into it, as well as the dependence of these processes on the structure of the metal, chemical, and crystal structure, stress–strain state, and other thermomechanical loads in the metal play an important role.

One of the most common models for describing hydrogen diffusion into metals is the trap model cf. [3, 4], which divides the hydrogen flows into a lattice flow inside the bulk and a hydrogen flow that is absorbed and accumulates in traps—dislocations, inhomogeneities, voids, microcracks, and other defects. The constants used in this model (such as the diffusion coefficient, probability of hydrogen jumping from normal lattice site to trap site, and probability of hydrogen releasing from trap site to normal lattice site) depend exponentially on the activation energy of hydrogen cf. [5]. Activation energy is a characteristic that describes the force of hydrogen interaction with the structural elements of a metal and its mobility inside a metal. This characteristic is determined experimentally for each material and is a table value.

Another type of model widely used to describe the hydrogen diffusion in metals is the probabilistic models cf. [6, 7]. They explain the absorption and diffusion of hydrogen with methods of statistical physics. They also use the hydrogen activation energy to take into account the impact of traps on the process of hydrogen transport from the external environment.

Multichannel diffusion models cf. [8, 9] describe the process using Fick's laws, but they also take into account the diffusion channels corresponding to trapped modes. They also use the value of the hydrogen activation energy.

Another motivation for a more detailed study of the hydrogen absorption and desorption process is the widespread method of long-term storage of hydrogen and its isotopes in chemical combination with titanium in the form of amorphous powder or solid samples cf. [10, 11]. Titanium absorbs hydrogen and its isotopes at relatively low temperatures (starting from 300 °C) and has sufficient capacity (experiments show that it can hold up to two hydrogen atoms per titanium atom). This long-term storage bases on the idea that the desorption of the stored hydrogen isotope occurs at sufficiently high temperatures when a chemical bond breaks or a trap filled with hydrogen is activated. However, sometimes the titanium and hydrogen isotope compounds decompose under normal conditions. Therefore this method needs to be investigated more precisely.

One can estimate the activation energy with various methods, for example, spectral methods cf. [12], or by measuring hydrogen permeability at different temperatures. But often these methods work adequately only for samples of specific shape or structure. For example, spectral methods give fairly accurate results only for the surface layer of the sample and hydrogen permeability can be measured only for thin samples in the form of films or membranes.

One of the most widespread methods for determining the activation energy of hydrogen in metal is the method of thermal desorption spectra cf. [13–15]. This experimental technique is based on the Kissinger approach cf. [16], which proposes one to consider the extraction of hydrogen from a sample as a first-order chemical reaction with activation energy equal to the binding energy. Using this method, one can adequately evaluate only surface sorption and desorption cf. [17], so to use this method one should neglect the hydrogen diffusion within the test sample. This contradicts the experimental data since taking diffusion into account, even in relatively small samples, introduces significant changes to the activation energy value cf. [18].

Moreover, some experiments show that activation energy value changes with a change in temperature or even during an isothermal process cf. [19]. This phenomenon is associated with the switching on of various types of traps and multichannel diffusion. Usually, in this case, the break in the Arrhenius dependences is ignored or considered as the sum of activation energies for each type of traps cf. [20].

All of the above factors lead to large variations in activation energy values for the same materials cf. [9, 21, 22].

In this work, we use the method of thermal desorption spectra to study samples of titanium hydrides and the desorption of hydrogen from them. We investigate amorphous powder samples. We investigate activation energy using the method proposed by Arrhenius and based on the data obtained, make assumptions about the nature of the desorption process and about diffusion models that are used to more accurately describe the process of absorption, desorption, and diffusion of hydrogen.

8.2 Experimental Technique

One of the methods for measuring the hydrogen concentration in a sample of a solid, which is used both in research and in industrial control, is the Vacuum Hot Extraction method (VHE) cf. [12, 23–25]. In this study, we obtain experimental data of industrial hydrogen analyzer AV-1 with the mass spectrometric recording of the time dependence of the hydrogen flow from the sample during its heating in vacuum. Samples are weighed on an analytical balance prior to measurements in a hydrogen analyzer.

Figure 8.1 (see Ref. [26]) shows a sample preparation system consisting of an extraction system 2, 3 (material—quartz glass) and a radiation furnace 4 with the constant temperature during the analysis. Cylindrical samples (1) placed in a cold pipe of an extractor (2) are made of thin quartz glass.

During the analysis, sample (1) dropped without breaking the vacuum into the analytical extractor pipe (3) with the constant extraction temperature automatically maintained by the heater (4) on the pipe. The functional diagram of the analyzer is shown in Fig. 8.2. The analyzer consists of an extractor, an infrared oven with a temperature controller, a vacuum system, a mass spectrometer, an analog-to-digital converter, and a computer with special software for working with the analyzer. The extractor consists of three connected pipes: a cold pipe for storing samples before analysis, a pipe for storing samples after analysis, and an analytical (hot) pipe. Samples can move from one pipe to another without breaking the vacuum.

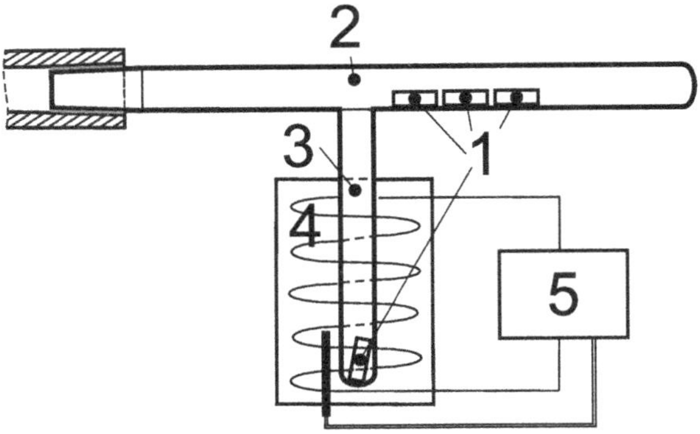

Fig. 8.1 Extraction system of hydrogen analyzer: 1—samples or container for powder and small particles , 2—extractor, 3—analytical scion, 4—radiation furnace with thermocouple, 5—furnace temperature controller

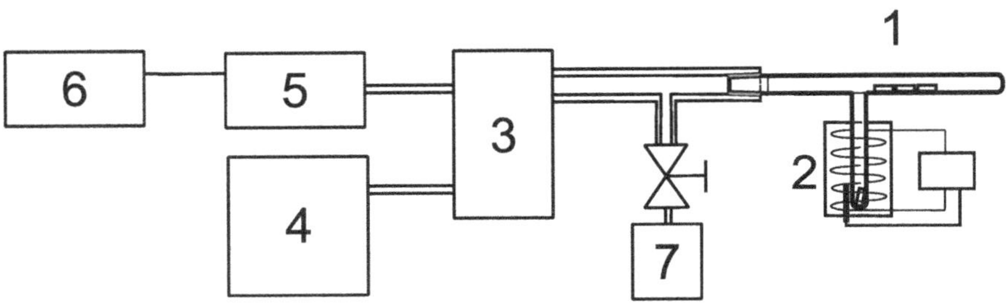

Fig. 8.2 Functional diagram of the AV-1 hydrogen analyzer: 1—extractor with samples; 2—radiation furnace with temperature controller, 3—nitrogen trap; 4—vacuum pump; 5—mass spectrometer; 6—digital registration system with a monitor; 7—system of gas calibration

When the hot pipe of the extractor is heated, hydrogen begins to flow out of its walls. These flows of hydrogen are called background flows. With further heating, the intensity of these flows continues to increase, and they may cause serious errors if the sample is analyzed while the extractor is heating. The hydrogen flows from the heated extractor itself may be comparable to or even exceed the hydrogen flow from the sample. Therefore, it is important to reduce the effect of background flows on the measurement (see work [27]). However, at a constant temperature, these background flows reach a stable level. To get the advantage of this during experiments, the test samples were first loaded into the cold scion of the extractor. Then the vacuum in the system is created. When the vacuum reaches the required level, the empty hot pipe of the extractor is kept at the extraction temperature until the background flows reach a stable level. Then the samples are loaded into the hot pipe of the extractor and held there until the hydrogen flow reaches the background level. This means that hydrogen flow from the sample is absent.

Taking into account the significant impact of the background flows, we apply a procedure of step increase of the extraction temperature. After the extraction system is prepared (the samples are loaded, the vacuum is pumped), the temperature is set below the temperature usually needed to extract all hydrogen from a sample (for most hydrides, it is in the range of 350–800 °C). The extractor volume is continuously pumped out by vacuum pumps of the analyzer to an operating pressure of 100 μPa. After the background flows settle, the analyzed sample moves into a hot pipe and warms up to the extraction temperature. The hydrogen flow evolved is measured by a mass spectrometric analyzer. After the hydrogen flow has reduced several times, we increase the temperature by a step and repeat the experiment. Measurements are finished when the hydrogen flux drops to background values.

As a result of analyzing the sample, the time dependence of the hydrogen flow from the extraction system (extraction curve) is obtained. The absolute values of the hydrogen flux are calculated using the calibration factor. The mass of hydrogen evolved is determined as an integral over time from the extraction curve relative to the background hydrogen flux.

Mass and volume concentrations are calculated using the initial mass of the sample and the mass of hydrogen extracted from the sample.

Pure iron containers diameter 7mm and 15mm high with a removable lid are used for samples of powder or fine particles. Container wall thickness is about 0.3 mm. The lid of the container is fixed not tightly to allow the evacuation of air from the container under vacuum and the extraction of hydrogen from the inside of the container. An empty container is degassed for an hour in vacuum at a temperature of 900 °C before being used for measurements.

We analyze titanium hydrides. The samples are in powder form with a particle size of 10–100 μm. All studied samples of titanium hydrides TiH_x were reference samples. Manufacturers indicated the hydrogen factor x in the accompanying documents. Some manufacturers have specified a range of $0.95 < x < 1.1$ After integrating the extraction curve, it turned out that some of them degraded before the beginning of the experiment during transportation and storage from TiH_2 up to $TiH_{0.3}$. However, such samples still can provide extraction curves.

8.3 Determination of the Activation Energy

To determine the activation energy, we use Kissinger's idea cf.[16] that the desorption of hydrogen from a metal can be considered as a chemical reaction described by the equation

$$\frac{dc}{dt} = k(1-c)^n \tag{8.1}$$

where c is the concentration of hydrogen, k is a rate constant, n is an order of the reaction. Although there is no certainty in which n to take, with sufficient accuracy one can assume that the reaction is of the first order. Thus, we can solve the equation and get the expression for the concentration:

$$c = 1 - A \exp^{-kt} \tag{8.2}$$

where A is some integration constant. If we denote hydrogen flow $\frac{dc}{dt}$ as q_H, from (8.1) and (8.2), we have

$$q_H = Ak \exp^{-kt} \tag{8.3}$$

and, having natural logarithm,

$$\ln q_H = \ln(Ak) - kt. \tag{8.4}$$

This formula can be used to generate a linear graph when processing experimental kinetic measurements. The dependence of hydrogen flow q_H on time t is graphically expressed by a straight line, and its slope is equal to $-k$. Thus, having the linear graph, we can easily calculate the rate constant k. To calculate the activation energy E_a from the rate constant, we use Arrhenius equation (see work [28])

$$k = k_0 \exp^{-\frac{E_a}{k_b T}} \tag{8.5}$$

where T is temperature and k_b is Boltzmann constant—if one takes it in eV, the activation energy is also in eV and is defined as energy per molecule. Taking the logarithm of both sides of the equation, we have

$$\ln k = \ln k_0 - \frac{E_a}{k_b T}. \tag{8.6}$$

Again, we generate a linear graph of the dependence of $\ln k$ on $1/T$ and find its slope that is equal to $-\frac{E_a}{k_b}$.

So to get the value of the activation energy from the dependence of hydrogen flow on time, one has first to find the scope of linear dependence of $\ln q_H$ on t, and second, find the scope of linear dependence of $\ln k$ on $1/T$.

8.4 Results and Discussion

We obtain extraction curves for the samples in form of powder. The summary of the experimental data is presented in Table 8.1.

Basically, all the extraction curves look very similar. A typical extraction curve is presented on Fig. 8.3. One can easily notice that the extraction curve has some typical regions: after the temperature is step-increased, it takes some time for the sample to heat, and then the hydrogen begins to eject—flow is decreasing exponentially. Then after sometime, the temperature is increased again, the flow jumps and begins to decrease again. This procedure is repeated several times.

Table 8.1 Experimental data

Type of compound	Temperature, °C	Integral over measurement time from background flow, ppm
TiH_2 -r	554, 590, 620, 690, 747, 900	2.1
TiH_2 -s	360, 370, 400	2.4
TiH	400, 450, 500, 900	2.25
TiH_2 -q	380, 390, 400	2.3
TiH_x	350	2.4
TiH_x	350	2.2
TiH_x	350	2.4

$_x$ Different samples of TiH_x ($x \approx 1$) were investigated

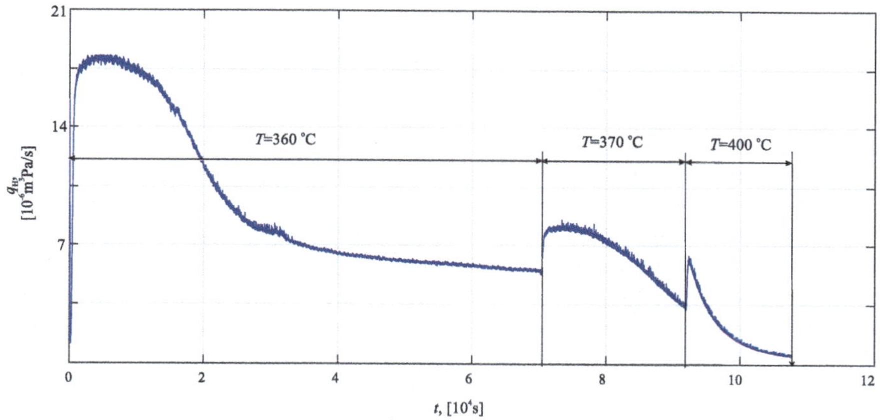

Fig. 8.3 Extraction curve for powder of TiH_2-s (see Table 8.1), temperatures are T =360 °C, 370 °C, 400 °C

First, we consider the Arrhenius dependence for TiH_2-r (first sample from Table 8.1). Red markers at Fig. 8.4 denote values obtained from the treatment of experimental data (extraction curve for TiH_2 at temperatures $T = 554$ °C, 590 °C, 620 °C, 690 °C, 747 °C, 900 °C). The scope of the linear dependence, multiplied by Boltzmann constant, is equal to the activation energy. In this case, we have two obvious linear dependences: $E_a = 1.880 \pm 0.375$ for lower temperatures $T = 554$ °C, 590 °C, 620 °C, 690 °C (blue dash and dots line) and $E_a = 0.136 \pm 0.033$ for higher temperatures $T = 747$ °C, 900 °C (green-dashed line).

This effect was already observed experimentally cf. [29–31], but still, there is no certainty why does the activation energy change. Authors of article [19] state that at lower temperatures various traps have more impact on the desorption process, than chemical bound at high temperatures, thus for the lower temperatures value of E_a is actually equal to $E_a^l = E_a + E_{traps}$, and for the high temperatures E_a has the table value of E_a. The impact of traps on the value of hydrogen is confirmed

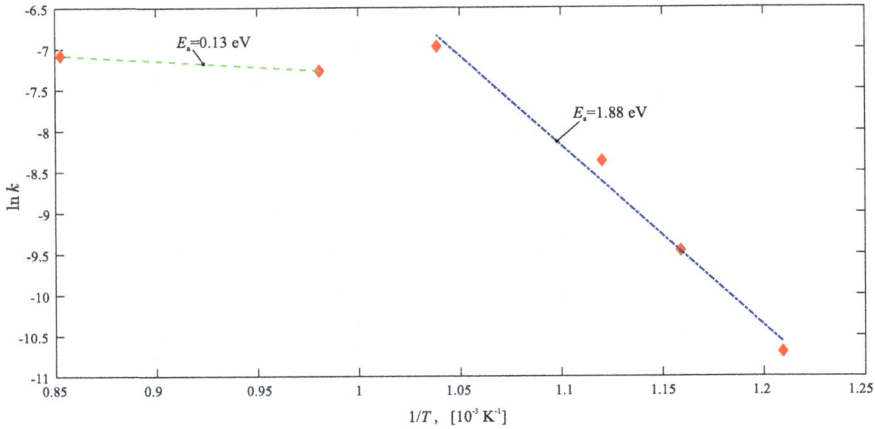

Fig. 8.4 Arrhenius dependencies for powdered samples of TiH_2-r (see Table 8.1), temperatures are $T = 554$ °C, 590 °C, 620 °C, 690 °C, 747 °C, 900 °C

by many researchers cf. [31–34]. They refer to the structure of the metal lattice, inhomogeneities, microcracks, and other defects or structure of the material.

These explanations cannot be implemented in our case, since we are investigating the desorption of a hydrogen from a powder, not from a sample with a metal lattice, so we can not have any bulk diffusion or most types of traps which change the value of the activation energy. In addition, the resulting activation energy values are much lower than the chemical bond energy (which is about 1–4 eV). Therefore, we conclude that our experiment cannot be described with Kissinger model as the chemical reaction of the first order. One of the explanations could be possible microchannels or voids in grains of powder particles, which store hydrogen and open at high temperatures, providing more intense extraction flow, but this assumption is a matter of future investigations.

In work [35], explanation of the two peaks on thermo-desorption spectra, and, subsequently, two values of $\ln k$ are explained as a result of skin effect. The concentration of hydrogen in metals at the boundary exceeds a concentration in the bulk more than ten times, forming a boundary layer of hydrogen. The extraction of metal with such hydrogen distribution gives two peaks at the thermo-desorption spectra for one temperature, that leads to two values of $\ln k$. One should notice that this explanation points out that titanium hydrides are not a chemical compound but a composite of the metal and the diffused hydrogen.

All the Arrhenius dependencies for the samples from the table are shown in Fig. 8.5. Different colors and markers (see Table 8.1) correspond to different samples—except for the black triangles, where three different experiments for TiH_x at temperature $T = 350$ °C are shown.

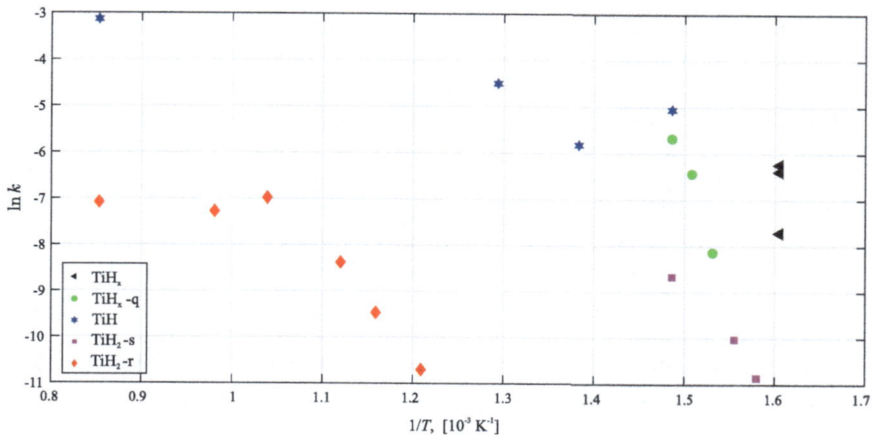

Fig. 8.5 Arrhenius dependences for the samples of TiH_2, TiH_x, TiH (see Table 8.1)

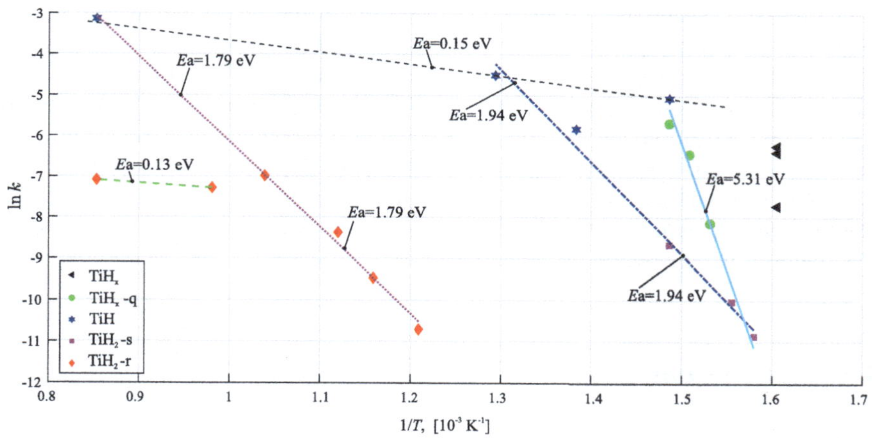

Fig. 8.6 Arrhenius dependences and linearisation for the samples of TiH_2, TiH_x, TiH (see Table 8.1)

From Fig. 8.5, it is clearly seen that the results fit well with linear dependencies even for different samples. Figure 8.6 shows the results for the linearization of the experimental points and the activation energies corresponding to these lines.

From Fig. 8.6, one can see that we have a tendency for the activation energy to decrease with the growth of temperature (or, conversely, increase at the range of low temperature). Although this effect again can be explained with the influence of some traps or inhomogeneities, we still do not have a clear idea of the processes occurring during the extraction and how they affect the value of the activation energy. In addition, we have some additional uncertainties.

We have two points at temperature of 900 °C (one for TiH_2 and another for TiH, these are most left dots at Fig. 8.6). The value of $\ln k$ is completely different at these points, and they fit different linear dependencies—even if we linearise our results in another way (for example, linearise results only for each sample). We have similar situation for TiH_2 at 360 °C and 370 °C—these two points fit well on two linear dependencies, although values of activation energies, corresponding to them, differ much (from $E_a = 1.94 \pm 0.505$ to $E_a = 5.309 \pm 1.108$ eV).

Values of activation energies at mid-range of temperatures (approximately 400–700 °C) have quite close values: $E_a = 1.94 \pm 0.505$ eV and $E_a = 1.794 \pm 0.431$ eV. But again next problem arises: for points at about 400 °C we have two values of activation energy, which, as it was mentioned above, differ very much.

Finally, we have three points at temperature 350 °C, all for TiH_x (black triangles at Fig. 8.5, Fig. 8.6). Although two points are close enough to be deviation of one point, the third has a distinct value of $\ln k$, which actually means that the value of activation energy at a temperature of 350 °C is close to infinity.

All these phenomenons need to be explained. One of the possible explanations is that mass of ejected hydrogen affects the value of the activation energy. As we mentioned before, despite our samples were standard titanium hydrides TiH_2 some of them degraded up to $TiH_{0.3}$. Nevertheless, if we consider the desorption process as the chemical reaction of the first order (according to Kissinger's model), the dependence of the activation energy on mass of the hydrogen should play any significant role. Again we conclude that explanation is that titanium hydrides are not the chemical compound but are a mixture or composite of titanium and hydrogen stored in inhomogeneities of titanium grains. The precise understanding of desorption process is a matter of the upcoming investigations.

8.5 Conclusions

To determine the binding energy of hydrogen in titanium hydride, we carried out various experiments under the same conditions for powder samples of titanium hydride. Based on the experimental data, we used the method of hot vacuum extraction of hydrogen to construct thermal desorption spectra at a stepwise change in temperature. Following Kissinger's model that the decomposition of a solid is a first-order chemical reaction, we built Arrhenius dependences, from which we determined the binding energy.

We found that the binding energy of hydrogen differs for different temperatures, and decreases with increasing temperature. This leads to the paradoxical result that lower binding energy hydrogen is extracted at high temperatures. Moreover, we found that for some temperatures the activation energy may differ, and for temperatures around 350–370 °C, the activation energy becomes equal to infinity. Since we have studied amorphous samples under the same conditions, we cannot explain these effects by various traps, inhomogeneities, and other properties of the material.

Furthermore, most of the obtained values of the activation energy (0.13–1.94 eV) do not correspond to the typical values of the chemical bond energy (2–4 eV).

One of the possible explanations is that titanium hydrides do not exist as a chemical compound, but are a composite of a metal and hydrogen diffused into it. This theory is supported by the results of work [35], which investigates the diffusion of hydrogen from samples, in which hydrogen is distributed in such a way that it forms a boundary layer. Diffusion with such a skin effect gives two peaks in the thermal desorption spectra.

Understanding the mechanism of hydrogen degradation in compounds with titanium is the subject of future research. This is important both for explaining the mechanisms of hydrogen embrittlement and for more efficient storage of hydrogen and its isotopes in compounds with titanium and other metals.

Acknowledgements The reported study was funded by RFBR, projects number 19-38-90298, 18-08-00201, 20-08-01100.

References

1. Cailletet, L., et al.: First report of H embrittlement of metals. Compt. Rend **58**, 327 (1864)
2. Johnson, W.H.: II. On some remarkable changes produced in iron and steel by the action of hydrogen and acids. Proc. Royal Soc. London **23**(156-163), 168–179 (1875)
3. McNabb, A., Foster, P.: A new analysis of diffusion of hydrogen in iron and ferritic steels. Trans. Metallurgical Soc. AIME **227**(3), 618 (1963)
4. Oriani, R.A.: The diffusion and trapping of hydrogen in steel. Acta Metallurgica **18**(1), 147–157 (1970)
5. Turnbull, A., Carroll, M., Ferriss, D.: Analysis of hydrogen diffusion and trapping in a 13% chromium martensitic stainless steel. Acta Metallurgica **37**(7), 2039–2046 (1989)
6. Kirchheim, R.: Hydrogen solubility and diffusivity in defective and amorphous metals. Progress Mater. Sci. **32**(4), 261–325 (1988)
7. Myers, S.M., Baskes, M., Birnbaum, H., Corbett, J.W., DeLeo, G., Estreicher, S., Haller, E.E., et al.: Hydrogen interactions with defects in crystalline solids. Rev. Modern Phys. **64**(2), 559 (1992)
8. Belyaev, A., Polyanskiy, A., Polyanskiy, V., Sommitsch, C., Yakovlev, Y.A.: Multichannel diffusion vs TDS model on example of energy spectra of bound hydrogen in 34CrNiMo6 steel after a typical heat treatment. Int. J. Hydrogen Energy **41**(20), 8627–8634 (2016)
9. Andronov, D.Y., Arseniev, D., Polyanskiy, A., Polyanskiy, V., Yakovlev, Y.A.: Application of multichannel diffusion model to analysis of hydrogen measurements in solid. Int. J. Hydrogen Energy **42**(1), 699–710 (2017)
10. Heung, L.: Titanium for long-term tritium storage. Tech. rep., Westinghouse Savannah River Co., Aiken, SC (United States) (1994)
11. Hayashi, T., Suzuki, T., Konishi, S., Yamanishi, T., Nishi, M., Kurita, K.: Development of ZrCo beds for ITER tritium storage and delivery. Fusion Sci. Technol. **41**(3P2), 801–804 (2002)
12. Nickel, N., Brendel, K., Saleh, R.: Laser crystallization of hydrogenated amorphous silicon. Physica Status Solidi **1**(5), 1154–1168 (2004)
13. Aveling, R.P.: Desorption spectrometers. US Patent 3-068-402 (1962)
14. Castro, F.J., Meyer, G.: Thermal desorption spectroscopy (TDS) method for hydrogen desorption characterization (i): theoretical aspects. J. Alloys Compounds **330**, 59–63 (2002)

15. Testi, M., Bartali, R., Crema, L.: Design and optimization of Isochoric Differential Apparatus (IDA) to reduce uncertainty in H2 sorption process measurements. Int. J. Hydrogen Energy (2020). https://doi.org/10.1016/j.ijhydene.2020.01.190
16. Kissinger, H.E.: Reaction kinetics in differential thermal analysis. Analyt. Chemistry **29**(11), 1702–1706 (1957)
17. Polyanskiy, A., Polyanskiy, V., Yakovlev, Y.A.: Experimental determination of parameters of multichannel hydrogen diffusion in solid probe. Int. J. Hydrogen Energy **39**(30), 17381–17390 (2014)
18. Padhy, G., Ramasubbu, V., Murugesan, N., Ramesh, C., Parvathavarthini, N., Albert, S.: Determination of apparent diffusivity of hydrogen in 9cr-1movnbn steel using hot extraction-PEMHS technique. Int. J. Hydrogen Energy **38**(25), 10683–10693 (2013)
19. Pisarev, A., Tsvetkov, I., Marenkov, E., Yarko, S. In: VII International School-Conference materials for young scientists and specialists IHISM'11: Interaction of hydrogen isotopes with structural materials (2011)
20. Pisarev, A., Ogorodnikova, O.: Elementary processes near the interface between hydrogen gas and solid. J. Nuclear Mater. **248**, 52–59 (1997)
21. Arseniev, D., Belyaev, A., Polyanskiy, A., Polyanskiy, V., Yakovlev, Y.: Benchmark study of measurements of hydrogen diffusion in metals. In: Dynamical Processes in Generalized Continua and Structures, pp. 37–61. Springer (2019)
22. Grigoriev, I., Meylichov, E.: Phizicheskie velichiny. The physical quantities) Energoizdat, Moscow (1991)
23. Petushkov, E., Tserfas, A., Maksumov, T.: Determination of hydrogen in molybdenum by a diffusion-manometric method. In: Arifov, U.A. (ed.) Secondary emission and structural properties of solids, pp. 107–109. Springer (1971)
24. Konar, J., Banerjee, N.: Vacuum heating hydrogen determination in aluminium and aluminium alloys. NML Techn. J. **16**(1–2), 18–19 (1974)
25. Sayi, Y., Ramakumar, K., Prasad, R., Yadav, C., Shankaran, P., Chhapru, G., Jain, H.: Determination of H2 and D2 content in metals and alloys using hot vacuum extraction. J. Radioanalyt. Nuclear Chemistry **230**(1–2), 5–9 (1998)
26. Kudinova, N.R., Polyanskiy, V.A., Polyanskiy, A.M., Yakovlev, Y.A.: Determining the bound energies of dissolved hydrogen on the basis of a multichannel diffusion model in a solid. St. Petersburg Polytechnical University Journal: Physics and Mathematics **1**(4), 347–355 (2015)
27. Zumberov, P., Kolesov, S., Polyansky, V., Varshavchik, E.: Calculating the activation energies of nickel, manufactured using 3D printing technology, with multichannel hydrogen diffusion model. In: E3S Web of Conferences, vol. 121, p. 4017–4021. EDP Sciences (2019)
28. Arrhenius, S.: Über die Dissociationswärme und den Einfluss der Temperatur auf den Dissociationsgrad der Elektrolyte (About the heat of dissociation and the influence of temperature on the degree of dissociation of the electrolytes). Zeitschrift für physikalische Chemie **4**(1), 96–116. De Gruyter Oldenbourg (1889)
29. Kehr K.: Theory of the diffusion of hydrogen in metals. In: Alefeld, G., Völkl, J. (eds.) Hydrogen in metals (I), pp. 197–226. Springer (1978)
30. Orimo, S., Matsushima, T., Fujii, H., Fukunaga, T., Majer, G.: Hydrogen desorption property of mechanically prepared nanostructured graphite. J. Appl. Physics **90**(3), 1545–1549 (2001)
31. Escobar, D.P., Verbeken, K., Duprez, L., Verhaege, M.: Evaluation of hydrogen trapping in high strength steels by thermal desorption spectroscopy. Mater. Sci. Eng. A **551**, 50–58 (2012)
32. Kirchheim, R.: Bulk diffusion-controlled thermal desorption spectroscopy with examples for hydrogen in iron. Metall. Mater. Trans. A **47**(2), 672–696 (2016)
33. Wei, F., Hara, T., Tsuzaki, K.: Precise determination of the activation energy for desorption of hydrogen in two Ti-added steels by a single thermal-desorption spectrum. Metall. Mater. Trans. B **35**(3), 587–597 (2004)
34. Ebihara, K.I., Kaburaki, H.: Numerical modeling of thermal desorption spectra of hydrogen: a review of thermal desorption models. ISIJ international **52**(2), 181–186 (2012)
35. Poyanskiy, V., Belyaev, A., Chevrychkina, A., Varshavchik, E.: Impact of skin efect of hydrogen charging on Choo-Lee diagramm for cylindrical samples (in press)

Chapter 9
Influence of Linear Elastic Stresses on Hydrogen Diffusion into Metals

Polina M. Grigoreva, Elena N. Vilchevskaya, and Vladimir A. Polyanskiy

Abstract Within the framework of linear nonequilibrium thermodynamics, we construct a new model of the diffusion of a gas component into a solid under thermomechanical loads. Assuming that we have a linear elastic behaviour of the solid, we obtain a local balance equation for the diffusion of the gas component, which takes into account the stress–strain state of the solid and its mutual influence on the diffusion process, the temperature in the system, and the concentration of the gas component infiltrated into the solid. We specify the model for the case of hydrogen diffusion into metal. The solution of the obtained differential equation shows that taking into account the stress–strain state strongly affects the distribution of hydrogen inside the metal. We found that the concentration quickly increases at the boundary layer, in which the hydrogen concentration exceeds the amount in bulk by more than a hundred times, which is consistent with experimental data on the skin effect when metals are saturated with hydrogen.

Keywords Hydrogen diffusion · Hydrogen embrittlement · Chemical potential · Mechanochemistry · Deformable media

9.1 Introduction

Hydrogen embrittlement of metals has been of scientific interest for more than a hundred years [1]. Many works, including those held in recent years, consider the critical local concentrations resulting from infiltration of hydrogen into metals from the aggressive corrosive environment. The data of Safronis [2] is recognized as the main source. Quantitative measurements of microstructural distributions of hydrogen

P. M. Grigoreva (✉) · E. N. Vilchevskaya · V. A. Polyanskiy
Institute for Problems in Mechanical Engineering RAS, Bolshoy pr., 61, V.O., St. Petersburg 199178, Russia
e-mail: polina.grigoreva239@gmail.com

V. A. Polyanskiy and A. K. Belyaev (eds.), *Advances in Hydrogen Embrittlement Study*, Advanced Structured Materials 143, https://doi.org/10.1007/978-3-030-66948-5_9

in metals are difficult. The Ag Decoration Method [3, 4] and other similar approaches cf. [5] provide only a qualitative map of the distribution of hydrogen. In the work of Safronis, the threshold local concentrations of hydrogen at which the influence of the HELP mechanism of hydrogen embrittlement begins is indicated as 0.1 H/M. This is equivalent to mass concentrations in 1500–9000 ppm range for different metals. However, this value is very high for most metals used nowadays. As technologies approach the extremely possible mechanical and chemical characteristics of metals and alloys, the maximum permissible concentration of hydrogen in modern alloys becomes about 50–100 times lower than in traditional [6]. Moreover, experiments based on mass spectrometry and electron microscopy also show that the presence of water vapour in the air, as well as aggressive environments, may serve as a source of hydrogen. Metals may absorb it, reducing their fracture toughness and strength [7].

Measurements of the hydrogen concentration profile in metals, both in aggressive environments and in the air [8–11] show that the hydrogen concentration at the boundary layer is tens or even hundreds of times higher than the concentration hydrogen in the entire volume of the metal, thereby creating some boundary layer. Recent experimental work [12–14] show that it is this thin layer of about 100 μm that has the main impact on the mechanical and strength properties of the metal.

Thus, modelling a thin boundary layer as a result of hydrogen diffusion from the environment and describing the mutual influence of the diffusion process on the mechanical properties of the metal is of considerable theoretical and practical interest.

There are some models of diffusion in solid bodies, which describe the ordinary forms of samples in a simple and stress–strain state. These models have a simplified design with only one specific mechanism of hydrogen redistribution. However, experiments show that hydrogen not only diffuses through a crystalline matter of metal but also redistributes in trap modes. This hydrogen redistribution proceeds according to several different variants [9, 15–17]. Therefore, the diffusion of hydrogen in metal should be considered as multichannel. Moreover, most authors use a trap model of diffusion (the fundamental works [18, 19] and many others, for example, [20–24]), where micro traps have different nature and characteristics of thermo-mechanical loads applied to the material. Such models are specific and do not adequately describe all the various experimental data. This leads to the fact that the diffusion coefficients of hydrogen in reference books differ by several orders of magnitude at the same temperature (for example, [25]).

One should also note that most trap and volumetric diffusion models were verified for small gradients of hydrogen concentration in the material. This does not allow one to take into account the almost hundred times difference in the levels of hydrogen concentration inside metal samples observed experimentally and requires substantial modification of these models.

The mechanism of local hydrogen plasticity was first described in [26] (HELP model). Later, in [2] and [27, 28], based on physical considerations about the potentials of the interaction of hydrogen with dislocations, the governing equations of the material were obtained. However, there are several uncertainties in this model. In particular, the authors neglect the nonlinear dependence of the internal potential on the magnitude of the stresses and hydrogen concentration, but at the same time, they consider large local hydrogen concentrations at which all nonlinearities should play a decisive role. It should be noted that only some metals such as zirconium allow saturation with hydrogen up to the HELP mechanism triggering concentrations of 0.1 H/M and higher without spontaneous destruction. Moreover, work [29] notes that the HELP model requires significant computational resources in solving any applied problem. The solution to this problem is to use a continuum model of the development of the dislocation. However, such a replacement is often inadequate. Therefore, the authors of the HELP model reduce all hydrogen problems in modelling the development of cracks and reducing crack resistance.

Recent publications that study the deformation and fracture of metals under mechanical stress in hydrogen-containing media [30–32] do not consider the process of hydrogen diffusion from the external environment. The models there were obtained heuristically and therefore cannot serve in the future as a basis for developing complex models which fully describe the process of hydrogen infiltration and diffusion.

Some works consider the hydrogen diffusion in metals which is not affected by any thermo-mechanical loads [33, 34]. They are focused on the influence of large plastic deformations on the redistribution of dissolved hydrogen [35–37], the influence of diffusively mobile hydrogen on the hydrogen embrittlement development [38, 39].

In this work, we make the first step of building a theoretical model to describe the heterogeneous distribution of the hydrogen and formation of a thin boundary layer in metal. Within the framework of linear nonequilibrium thermodynamics, we obtain a modified local equation of balance of the diffusion component. This equation takes into account the mutual influence of diffusion and the stress–strain state, as well as the dependence of the diffusion process on temperature, gas concentration and other thermomechanical loads.

As the first attempt to start the description of hydrogen diffusion in metals and see how that approach works, we restrict ourselves to an isotropic linear continuum for the metal and abandon the diffusion of hydrogen in trap modes. The first assumption corresponds to the case when hydrogen concentrations are not large enough to cause plastic deformations. The assumption that there are no hydrogen traps in our model will not considerably affect the results since it is known mostly hydrogen with low binding energy (i.e. diffusively mobile hydrogen) affects the stress–strain state of metal [40] and vice versa, zones of tensile stresses 'attract' low-binding hydrogen (the so-called Gorsky effect, [41]).

The constructed model is verified on the boundary problem of determining the stress-strain state of a cylindrical sample under uniaxial tension and the distribution of gas concentration. We use the results to estimate the possibility of further complicating the model to describe the formation of the experimentally observed boundary layer.

9.2 Problem Statement and Governing Equations

9.2.1 Chemical Potential

One of the most common ways of describing the diffusion process, taking into account the effect of temperature, heat fluxes, the local concentration of the diffusion component and the stress-strain state of a solid in which diffusion occurs, is the theory of chemical potential. Introduced by Onsager and Prigogine [42, 43] for multicomponent mixtures of diffusing gases, it has been used for solids, as well as for various diffusion systems under different thermo-mechanical loads.

Within the framework of linear nonequilibrium thermodynamics, one may consider various thermo-mechanical loads which affect a system of a diffusing gas in a solid. The chemical potential can include the influence of a stress-strain state of solid, plastic deformations, heat fluxes in the system, trapping modes and other effects that influence the diffusion process. One can note that, although some phenomena cannot be introduced non-phenomenologically (for example, plasticity or the influence of trapping modes), the theory of chemical potential is the only theory that allows one to take into account all these effects within the framework of one approach. This makes it possible not to introduce additional restrictions on the range of external conditions to keep the model corresponding to the experiment. Moreover, even the phenomenological account of any effect in the chemical potential is carried out following the basic principles of mechanics and thermodynamics. For example, the effect of plasticity can be considered through the deformations (using the decomposition of the deformation gradient into elastic and plastic parts) or by introducing an additional term into the bulk energy of a solid and taking into account the dissipation of this energy under plastic effects.

In this work, we verify the basic principles of the theory of chemical potentials and consider how diffusion is affected by small linear elastic deformations in a solid material, as well as how strongly these deformations affect the diffusion process. We also want to figure out whether they can explain the skin effect, observed in the diffusion of hydrogen into metals from the environment. In this work, we neglect the phenomena of plasticity, trap modes and other phenomena.

Following the non-equilibrium linear thermodynamic approach, we define diffusion flux j as

$$j = \frac{Dc}{RT}\nabla\mu, \tag{9.1}$$

where D is a diffusivity constant, c is the concentration, R is the gas constant, T is a temperature and μ is the chemical potential of the diffusing gas. The chemical potential of the ideal gas in vacuum is equal to

$$\mu_g = \mu_0(T) + RT \ln \tfrac{c}{c_*}. \tag{9.2}$$

Here, c_* is some concentration of normalization and μ_0—the reference chemical potential.

The chemical potential of the diffusing constituent consists of the chemical potential of the gas itself, as well as of other terms that affect the diffusion process. They enter the Eq. (9.1) as terms in general chemical potential of the diffusing component. For now, we consider only the influence of the stress–strain state of the solid on the diffusion.

We will consider the chemical potential of the solid in its full form [44, 45], since taking into account only the first term of expansion or term, associated only with pressure (as it is done in a lot of works, for example, [46]) imposes strong restrictions on the range of external conditions. The full chemical potential of a solid is defined as the (canonical) Eshelby stress tensor

$$\boldsymbol{\mu}_s = M\left(f\boldsymbol{E} - \frac{1}{\rho}\boldsymbol{F}^T \cdot \boldsymbol{S}\right), \tag{9.3}$$

where M is a molar mass of a solid, ρ is its density, f is a density of a bulk energy ($f = w/\rho$), $\boldsymbol{F}$ is a deformation gradient and $\boldsymbol{S}$ is a Piola-Kirchhoff stress tensor.

Since we do not have any phase transitions or moving boundary, the diffusion flux, induced by the stress-strain state of the solid, is proportional to the trace of the chemical potential tensor [47, 48]. The full diffusion flux is equal to

$$j = \frac{Dc}{RT}\nabla\mu = \frac{Dc}{RT}\nabla\left(\mu_g + tr\boldsymbol{\mu}_s\right). \tag{9.4}$$

9.2.2 *Case of Linear-Elastic Strains*

In this work, we consider the linear elastic behaviour of the solid. We do not solve any heat equations and consider temperature T as a constant, and neglect all effects of plasticity and trapping.

In the case of linear elastic strain,

$$w = \eta(T) + \tfrac{1}{2}\boldsymbol{\varepsilon}_e \cdot\cdot\, \boldsymbol{C} \cdot\cdot\, \boldsymbol{\varepsilon}_e, \tag{9.5}$$

where η is the free energy volume density of the stress-free solid (so-called chemical energy) and depends only on temperature.

Thus, we can write an expression for the $\mu_s = tr\boldsymbol{\mu}_s$:

$$\mu_s = \frac{M}{\rho}\left(3\eta(T) + \tfrac{3}{2}\boldsymbol{\varepsilon}_e \cdot\cdot\, \boldsymbol{C} \cdot\cdot\, \boldsymbol{\varepsilon}_e - \boldsymbol{\sigma} \cdot\cdot\, \boldsymbol{\varepsilon}_e\right) \tag{9.6}$$

Here, we denote elastic strains as $\boldsymbol{\varepsilon}_e$, $\boldsymbol{\sigma}$ is the Cauchy stress tensor and $\boldsymbol{C}$ is the stiffness tensor. We set $\boldsymbol{\varepsilon}_*$ for strains, induced by the diffused gas into the solid, $\boldsymbol{\varepsilon}_* = \boldsymbol{\varepsilon}_*(c)$. Thus, following Hooke's law for linear elastic behaviour of material, we obtain

$$\mu_s = \frac{M}{\rho}\left[3\eta(T) + \tfrac{1}{2}\left(\boldsymbol{\varepsilon} - \boldsymbol{\varepsilon}_*\right)\cdot\cdot\,\boldsymbol{C}\cdot\cdot\left(\boldsymbol{\varepsilon} - \boldsymbol{\varepsilon}_*\right)\right]. \tag{9.7}$$

As $\eta(T)$ does not depend on coordinate and we consider that change of density of solid body ρ is insignificant, for the gradient of chemical potential, induced by solid body, finally we have that

$$\nabla\mu_s = \frac{M}{2\rho}\nabla\left[\left(\boldsymbol{\varepsilon} - \boldsymbol{\varepsilon}_*\right)\cdot\cdot\,\boldsymbol{C}\cdot\cdot\left(\boldsymbol{\varepsilon} - \boldsymbol{\varepsilon}_*\right)\right] \tag{9.8}$$

As we consider our body as isotropic, we use stiffness tensor with two constants, Young's modulus E and Poisson's ratio ν. We also follow the ideas from the problems of thermal elasticity and set deformation, induced by the diffused gas component, proportional to the concentration: $\boldsymbol{\varepsilon}_* = \alpha c\boldsymbol{E}$. So, we obtain

$$\nabla\mu_s = \frac{E}{2\rho(1+\nu)}\nabla\left[\frac{\nu}{1-2\nu}\left(tr\boldsymbol{\varepsilon} - 3\alpha c\right)^2 + \boldsymbol{\varepsilon}\cdot\cdot\boldsymbol{\varepsilon} - 2\alpha c\,tr\boldsymbol{\varepsilon} + 3\alpha^2c^2\right]. \tag{9.9}$$

Finally, we recall that $\nabla\cdot\boldsymbol{j} = \frac{\partial c}{\partial t}$ and substitute obtained expressions for the chemical potential into the (9.4). After opening brackets, we get the diffusion equation:

$$\frac{\partial c}{\partial t} = \nabla\cdot\left(D_{\text{eff}}\nabla c + Vc\right), \tag{9.10}$$

where

$$D_{\text{eff}} = D_0\left(1 + \frac{cM}{RT\rho}\frac{E}{1-2\nu}\left[-\alpha tr\boldsymbol{\varepsilon} + 3\alpha^2 c\right]\right) \tag{9.11}$$

$$V = \frac{D_0 ME}{RT\rho}\left(\frac{\nu}{(1+\nu)(1-2\nu)}tr\boldsymbol{\varepsilon}\nabla(tr\boldsymbol{\varepsilon}) - \frac{1}{1-2\nu}\alpha c\nabla(tr\boldsymbol{\varepsilon}) + \frac{1}{2(1+\nu)}\nabla(\boldsymbol{\varepsilon}\cdot\cdot\boldsymbol{\varepsilon})\right)$$

In this new diffusion equation, the coefficient at the concentration gradient (diffusion coefficient) is not a constant value. It depends on the stress–strain state of solid, elastic constants, the concentration of the gas component in the solid (metal) and the temperature in the system. In the diffusion equation, we also obtain an additional term which is proportional to the gas concentration. The proportionality coefficient also depends on the thermo-mechanical loads in the system. This term is a force that decelerates the diffusion and rises when the concentration in the solid increases. One should also notice that all the deformations have an influence on the diffusion process, and therefore approximation with the first term of decomposition and spherical part of the Cauchy stress tensor is not relevant.

We can also note that the resulting equation does not impose any restrictions on the values of temperatures, deformations, elastic moduli, concentration values and gradients of concentration and deformations inside the body, except that we take such a deformation field to remain within the framework of the linear theory of elasticity.

9.3 Boundary-Value Problem for Axially-Symmetric Body

Since most of the experiments are performed on cylindrical samples, in which tensile stresses are applied along the axis of symmetry, we consider the boundary value problem for a cylinder under uniaxial tension. Suppose it is located so that the symmetry axis is directed along the $0z$ axis, and we can define the cylinder as $z \in [-h, h]$; $r \in [0, r_o]$ (Fig. 9.1). We believe that its linear dimensions with respect to z significantly exceed the radius r_o, and therefore, the cylindrical sample is in a plane-deformed state before the diffusion process begins. We consider that $u_r = u_r(r)$ and does not depend on z-coordinate.

We also assume that the hydrogen infiltrates only from the side surface, and due to the stress–strain state and the symmetry concentration depends only on the r-coordinate, $c = c(r)$.

The elasticity equation in cylindrical coordinates will look as follows:

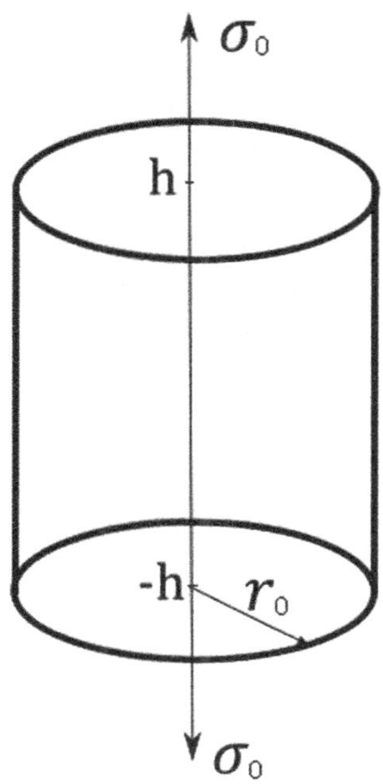

Fig. 9.1 The cylindrical sample under axial tension; the hydrogen diffuses through the outer boundary of the sample at $r = r_o$

$$\frac{\partial \sigma_r}{\partial r} + \frac{\sigma_r - \sigma_\varphi}{r} + \frac{\partial \tau_{rz}}{\partial z} = 0 \qquad (9.12)$$
$$\frac{\partial \sigma_z}{\partial z} + \frac{\tau_{rz}}{r} + \frac{\partial \tau_{rz}}{\partial r} = 0$$

Here, σ_r, σ_φ and τ_{rz} are the components of the Cauchy stress tensor:

$$\sigma_r = \frac{E}{1+\nu}\left[\frac{\nu}{1-2\nu}\left(\frac{\partial u_r}{\partial r} + \frac{u_r}{r} + \frac{\partial u_z}{\partial z} - 3\alpha c\right) + \frac{\partial u_r}{\partial r} - \alpha c\right]$$
$$\sigma_\varphi = \frac{E}{1+\nu}\left[\frac{\nu}{1-2\nu}\left(\frac{\partial u_r}{\partial r} + \frac{u_r}{r} + \frac{\partial u_z}{\partial z} - 3\alpha c\right) + \frac{u_r}{r} - \alpha c\right]$$
$$\sigma_r = \frac{E}{1+\nu}\left[\frac{\nu}{1-2\nu}\left(\frac{\partial u_r}{\partial r} + \frac{u_r}{r} + \frac{\partial u_z}{\partial z} - 3\alpha c\right) + \frac{\partial u_z}{\partial z} - \alpha c\right] \qquad (9.13)$$
$$\tau_{rz} = \frac{E}{1+\nu}\frac{\partial u_z}{\partial r}$$

The boundary conditions are prescribed tensile uniaxial stress and stress-free side surface:

$$u_r|_{r=0} < \infty; \qquad \sigma_r|_{r=r_o} = 0 \qquad (9.14)$$
$$\sigma_z|_{z=\pm h} = \sigma_0.$$

The diffusion equation will look as follows:

$$\frac{\partial c}{\partial t} = \frac{\partial}{\partial t}\left(D_{\text{eff}}\frac{\partial c}{\partial r} + Vc\right) \qquad (9.15)$$

where

$$D_{\text{eff}} = D_0\left(1 + \frac{cM}{RT\rho}\frac{E}{1-2\nu}\left[-\alpha\left(\frac{\partial u_r}{\partial r} + \frac{u_r}{r} + \frac{\partial u_z}{\partial z}\right) + 3\alpha^2 c\right]\right) \qquad (9.16)$$

$$V = \frac{D_0 M E}{RT\rho}\left[\frac{\nu}{(1+\nu)(1-2\nu)}\left(\frac{\partial u_r}{\partial r} + \frac{u_r}{r} + \frac{\partial u_z}{\partial z}\right)\left(\frac{\partial^2 u_r}{\partial r^2} - \frac{u_r}{r^2} + \frac{1}{r}\frac{\partial u_r}{\partial r} + \frac{\partial^2 u_z}{\partial z \partial r}\right)\right.$$
$$- \frac{1}{1-2\nu}\alpha c\left(\frac{\partial^2 u_r}{\partial r^2} - \frac{u_r}{r^2} + \frac{1}{r}\frac{\partial u_r}{\partial r} + \frac{\partial^2 u_z}{\partial z \partial r}\right) \qquad (9.17)$$
$$\left. + \frac{1}{2(1+\nu)}\left(\frac{u_r}{r^2}\frac{\partial u_r}{\partial r} - \frac{u_r^2}{r^4} + \frac{\partial u_r}{\partial r}\frac{\partial^2 u_r}{\partial r^2}\right)\right]$$

The boundary condition is the prescribed concentration at the outer surface of the body:

$$c|_{r=0} < \infty; \quad c|_{r=r_o} = c_* \tag{9.18}$$

The initial condition is the absence of the concentration in the body:

$$c|_{t=0} = 0. \tag{9.19}$$

One should note that we have terms that depend on the concentration in the elastic problem and the coefficients that depend on deformations (both in r- and in z-directions) in the diffusion equation. Thus, we have the coupled problem, and these two equations cannot be solved separately.

9.4 Numerical Solution and Discussion

The boundary problem (both elasticity problem and diffusion equation) is solved numerically using the explicit finite difference scheme, which corresponds to the finite volume method. Experimental data is taken from [49] for steel T24. Parameter α corresponding to the expansion due to the hydrogen diffusion is chosen to keep deformations in the linear elastic area. All parameters are presented in Table 9.1.

Since we consider the capabilities of the presented model, we can neglect possible inaccuracies or inconsistencies in the parameters and consider them as parameters of some model material.

We were able to perform calculations only for the timeless than 6 h: after this time the stresses near the boundary layer start to exceed linear elasticity and the stress–strain state of the steel should be described within the plasticity theory. Due to these restrictions, we investigate only the beginning of the diffusion process.

The two main parameters governing the diffusion process are the effective diffusivity coefficient D_{eff} and the coefficient at the linear concentration term V. They both depend on the hydrogen concentration and the deformation values.

The value of the diffusivity coefficient significantly depends on the hydrogen concentration and the stress–strain state of the steel, since the value of the effective

Table 9.1 Material and geometry parameters value [49]

Parameters	Value
Diffusivity coefficient D_0, mm^2/s	0.035
Temperature T, K	293
Young's modulus E, GPa	182
Poisson's ratio ν	0.295
ρ/M,mol/m^3	14.5e3
Outer radius r_o, mm	1.1
Expansion coefficient α	0.03

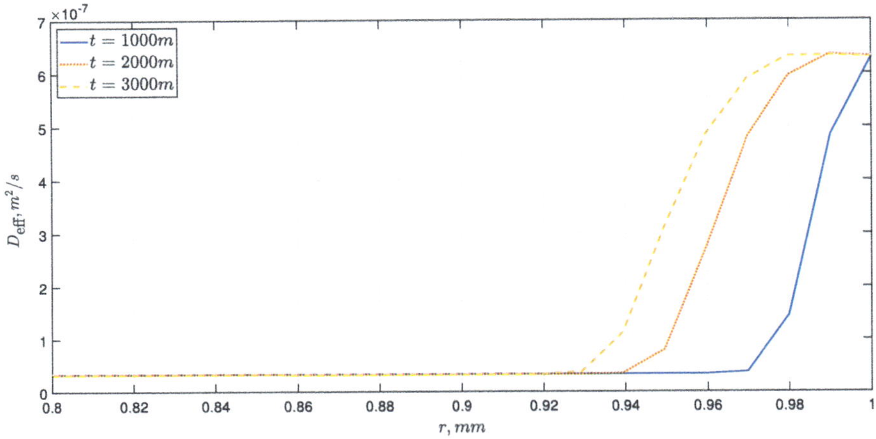

Fig. 9.2 Dependence of diffusivity coefficient D_{eff} on coordinate at different times, $\sigma_0 = 0.1$ GPa

diffusivity coefficient differs within the sample by almost ten times (see Fig. 9.2). It can also be noted that the effective diffusion coefficient increases with time and the rate of its growth is variable and decreases with time. The profile of the effective diffusion coefficient dependence on the coordinate practically does not change.

The concentration coefficient V in the second term of the (9.10) equation is very small (of the order of 1e5) and does not vary much with time, thus practically without exerting any influence on the diffusion process. In further studies of the model in the theory of small deformations, this term can be neglected. It should be noted that the real deformations arising from expansion due to the hydrogen infiltration into the steel are not small and exceed 3%, therefore, we can say that the coefficient V begins to influence the diffusion process at large finite deformations and significant gradients of them.

Although at short times hydrogen infiltrates into the metal rather quickly, from Fig. 9.3 we see that after some time the process of hydrogen redistribution slows down, and then with increasing time the distribution of hydrogen concentrations changes insignificantly. We can also notice that yet the diffusivity coefficient is still large at this boundary zone, the diffusivity coefficient in the bulk is significantly smaller (Fig. 9.2). This leads to the higher saturation with hydrogen of the boundary layer and not the propagation of the concentration profile deeper.

This experimentally observed effect is usually associated with the filling of hydrogen traps inside the crystal lattice of the metal. However, the experimentally observed multiple difference between the hydrogen concentration in the central and near-boundary parts of the samples is often inexplicable, since the experimental samples are made so that mechanical treatment does not violate the initial homogeneity of the metal throughout the volume of the sample. In our model, we obtain the blocking effect at saturation with hydrogen from the general propositions of nonlinear

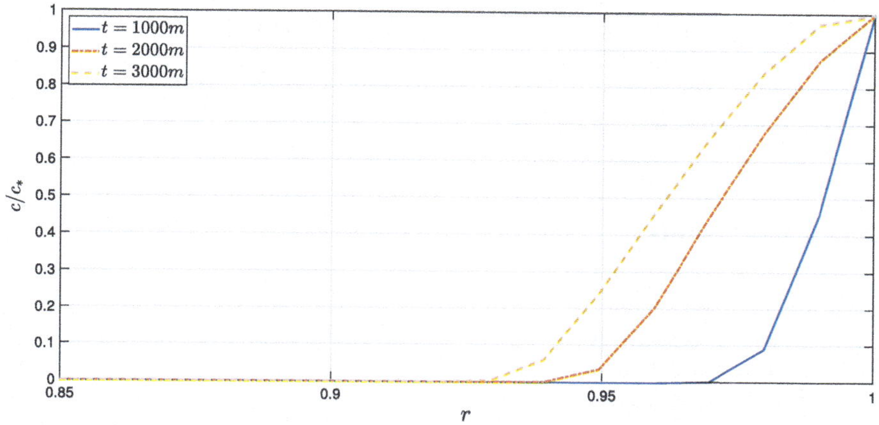

Fig. 9.3 Dependence of normalized concentration c/c_* on coordinate at different times, $\sigma_0 = 0.1$ GPa

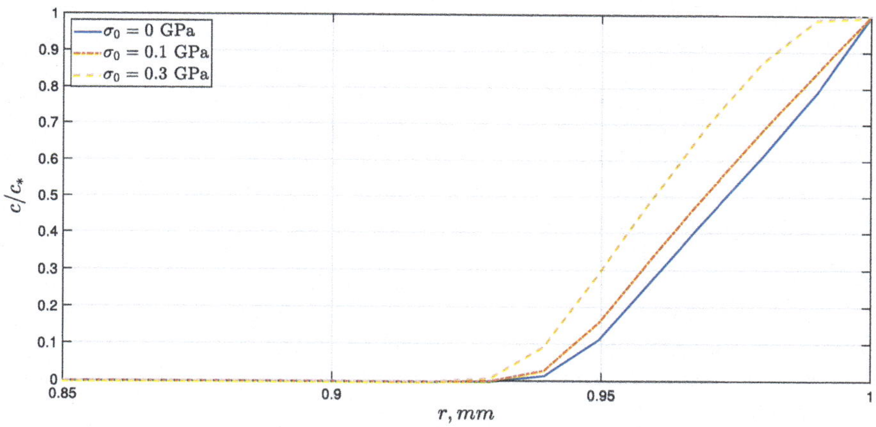

Fig. 9.4 Dependence of normalized concentration c/c_* on coordinate at different axial stresses σ_0, $t = 2000$ m

nonequilibrium thermodynamics, which was previously described only with the help of empirical or phenomenological terms introduced into the diffusion equation.

The obtained concentration profile inside the metal is highly nonuniform: the concentration values near the boundary and in a thin boundary layer are hundreds of times higher than the concentration inside the sample. This indicates the agreement of the model with the experiment and describes the experimentally observed boundary layer as part of the skin effect.

An increase in tensile axial stresses does not significantly change the diffusion process and the concentration profile (Fig. 9.4). With an increase in tensile stresses, diffusion proceeds faster, accelerating significantly at stresses close to the yield point.

This is due to both the fact that the crystal lattice of the metal expands, allowing hydrogen molecules to diffuse into the metal with less resistance, and with the fact that stretching along the axis of symmetry reduces the value of the locking stresses, thereby weakening the locking effect.

9.5 Conclusions

We have built a new model of the diffusion of the gas component into a solid based on the theory of chemical potentials and the principles of linear nonequilibrium thermodynamics. We took the trace of the Eshelby energy–momentum tensor in its full form as the chemical potential of a solid body. To evaluate the effect of deformations in a solid, we considered the case of linear elasticity and did not consider plasticity, heat effects, and large deformations. We also took into account the deformations that arise as a result of hydrogen diffusion inside the solid and introduced the spherical tensor of diffusion deformations, which is linearly proportional to the concentration.

As a result, we obtained an equation for the local balance of the diffusion component, which takes into account the dependence of the diffusion process on the stress-strain state of the solid, the concentration of the gas component inside it, and other thermo-mechanical loads. The new diffusion equation is a modified Fick's equation with a non-constant diffusivity coefficient and an additional term in the expression for the diffusion flow, which is linearly proportional to the concentration of the gas component, and the proportionality coefficient also depends on the deformations in the solid and the concentration of the gas component.

Using the finite volume method, we obtained a numerical solution of the coupled boundary value problem of determining the stress–strain state of a cylindrical steel sample under uniaxial tension and the distribution of diffused hydrogen concentration. We found that hydrogen quickly saturates a thin boundary zone, and with time the concentration profile only moves deeper practically unchanged into the sample.

As a result of hydrogen diffusion into the steel, large deformations arise inside the sample. They significantly increase the diffusion coefficient. Thus, saturation of the thin layer occurs faster, than further propagation of the concentration profile into the sample. This locking effect is also observed experimentally. Earlier in the literature, this locking effect was explained by the presence of traps which were phenomenologically introduced into the model. We describe this locking effect from the first principles.

In addition, we found that in the case of small linear deformations, the term proportional to the concentration is small and practically does not change with time, which suggests that the effect of this term in the case of small deformations can be neglected. This result can be used with further complications of the model; however, in the case of large strains and large strain gradients, this term is likely to have a significant effect.

To study the diffusion process completely, it is necessary to add plasticity to the model, since deformations corresponding to plastic deformations arise in the sample with further saturation with hydrogen. In this work, we investigated diffusion at short times (up to 6 h).

The resulting diffusion model is the first step to build a complete model of hydrogen diffusion into metals from the environment. Further improvements of the model can be taking into account plastic deformations, metal inhomogeneities and associated internal stresses.

Acknowledgements The work was supported by the Russian Science Foundation (RCF), grant number 18-19-00160.

References

1. Cailletet, L., et al.: First report of H embrittlement of metals. Compt. Rend **58**, 327 (1864)
2. Sofronis, P., Liang, Y., Aravas, N.: Hydrogen induced shear localization of the plastic flow in metals and alloys. Eur. J. Mech.-A/Solids **20**(6), 857–872 (2001)
3. Ronevich, J., Spee,r J., Krauss G., Matlock D.: Improvement of the hydrogen microprint technique on AHSS steels. Metallogr. Microstruct. Analys.**1**(2), 79–84 (2012)
4. Akiyama, E., Matsuoka, S.: Hydrogen visualization in steels using Ag decoration method. Mater. Trans. **56**(6), 793–797 (2015)
5. Evers, S., Senöz, C., Rohwerder, M.: Hydrogen detection in metals: a review and introduction of a Kelvin probe approach. Sci. Technol. Adv. Mater. **14**(1), 14201 (2013)
6. Nie, Y., Kimura, Y., Inoue, T., Yin, F., Akiyama, E., Tsuzaki, K.: Hydrogen embrittlement of a 1500-MPa tensile strength level steel with an ultrafine elongated grain structure. Metallurg. Mater. Trans.A **43**(5), 1670–1687 (2012)
7. Khrustalev, Y.A., Simakov, Y.S., Glazunov, M., Gubin, V.: Formation of hydrogen under the metal friction. Zhurnal Fizicheskoi Khimii **63**(5), 1355–1357 (1989)
8. Frolova, K., Vilchevskaya, E., Polyanskiy, V., Alekseeva, E.: Modelling of a hydrogen saturated layer within the micropolar approach. In: Abali, B., Altenbach, H., dell'Isola, F., Eremeyev, V., Öchsner, A. (eds.) New Achievements in Continuum Mechanics and Thermodynamics, vol. 108, pp. 117–128. Springer, Cham (2019)
9. López-Suárez, A., Valencia, C.E., López-Patiño, J., Vargas, M.C., Fuentes, B.E.: Improvement of titanium hydrogenation by low energy ion irradiation. Int. J. Hydrogen Energy **40**(11), 4194–4199 (2015)
10. Wu, T.I., Wu, J.C.: Effects of cathodic charging and subsequent solution treating parameters on the hydrogen redistribution and surface hardening of Ti-6Al-4V alloy. J. Alloys Compounds **466**(1–2), 153–159 (2008)
11. Martinsson, Åsa, Sandström, R.: Hydrogen depth profile in phosphorus-doped, oxygen-free copper after cathodic charging. J. Mater. Sci.**47**(19), 6768–6776 (2012)
12. Arseniev, D., Belyaev, A., Polyanskiy, A., Polyanskiy, V., Yakovlev, Y.: Benchmark study of measurements of hydrogen diffusion in metals. In: Dynamical Processes in Generalized Continua and Structures, pp. 37–61. Springer (2019)
13. Polyanskiy, V., Belyaev, A., Alekseeva, E., Polyanskiy, A., Tretyakov, D., Yakovlev, Y.A.: Phenomenon of skin effect in metals due to hydrogen absorption. Continuum Mechan. Thermodyn. **31**(6), 1961–1975 (2019)
14. Andronov, D.Y., Arseniev, D., Polyanskiy, A., Polyanskiy, V., Yakovlev, Y.A.: Application of multichannel diffusion model to analysis of hydrogen measurements in solid. Int. J. Hydrogen Energy **42**(1), 699–710 (2017)

15. Hadam, U., Zakroczymski, T.: Absorption of hydrogen in tensile strained iron and high-carbon steel studied by electrochemical permeation and desorption techniques. Int. J. Hydrogen Energy **34**(5), 2449–2459 (2009)
16. Turnbull, A.: Perspectives on hydrogen uptake, diffusion and trapping. Int. J. Hydrogen Energy **40**(47), 16961–16970 (2015)
17. Liu, Q., Venezuela, J., Zhang, M., Zhou, Q., Atrens, A.: Hydrogen trapping in some advanced high strength steels. Corrosion Sci. **111**, 770–785 (2016)
18. Oriani, R.A.: The diffusion and trapping of hydrogen in steel. Acta Metallurgica **18**(1), 147–157 (1970)
19. McNabb, A., Foster, P.: A new analysis of diffusion of hydrogen in iron and ferritic steels. Trans. Metallurg. Soc. AIME **227**(3), 618 (1963)
20. Oudriss, A., Creus, J., Bouhattate, J., Conforto, E., Berziou, C., Savall, C., Feaugas, X.: Grain size and grain-boundary effects on diffusion and trapping of hydrogen in pure nickel. Acta Materialia **60**(19), 6814–6828 (2012)
21. Toribio, J., Kharin, V.: A generalised model of hydrogen diffusion in metals with multiple trap types. Philos. Magaz. **95**(31), 3429–3451 (2015)
22. Takahashi, J., Kawakami, K., Kobayashi, Y.: Origin of hydrogen trapping site in vanadium carbide precipitation strengthening steel. Acta Materialia **153**, 193–204 (2018)
23. Depover, T., Van den Eeckhout, E., Verbeken, K.: Hydrogen induced mechanical degradation in tungsten alloyed steels. Mater. Characterization **136**, 84–93 (2018)
24. Zhang, Z., Moore, K.L., McMahon, G., Morana, R., Preuss, M.: On the role of precipitates in hydrogen trapping and hydrogen embrittlement of a nickel-based superalloy. Corrosion Sci. **146**, 58–69 (2019)
25. Grigoriev, I., Meylichov, E.: Phizicheskie velichiny (The physical quantities). Energoizdat, Moscow (1991)
26. Birnbaum, H.K., Sofronis, P.: Hydrogen-enhanced localized plasticity—a mechanism for hydrogen-related fracture. Mater. Sci. Eng.: A **176**(1–2), 191–202 (1994)
27. Delafosse, D., and Magnin, T.: Interfaces in stress corrosion cracking: a case study in duplex stainless steels. In: Solid State Phenomena, vol. 59, pp. 221–250. Trans Tech Publ (1998)
28. Delafosse, D., Magnin, T.: Hydrogen induced plasticity in stress corrosion cracking of engineering systems. Eng. Fracture Mech. **68**(6), 693–729 (2001)
29. Ignatenko, A., Pokhodnya, I., Paltsevich, A., Sinyuk, V.: Dislocation model of hydrogen-enhanced localizing of plasticity in metals with BBC latttice. Paton Welding J. **3**, 15–19 (2012)
30. Yamabe, J., Yoshikawa, M., Matsunaga, H., Matsuoka, S.: Hydrogen trapping and fatigue crack growth property of low-carbon steel in hydrogen-gas environment. Int. J. Fatigue **102**, 202–213 (2017)
31. Yamabe, J., Yoshikawa, M., Matsunaga, H., Matsuoka, S.: Effects of hydrogen pressure, test frequency and test temperature on fatigue crack growth properties of low-carbon steel in gaseous hydrogen. Procedia Structural Integrity **2**, 525–532 (2016)
32. Yamabe, J., Takakuwa, O., Matsunaga, H., Itoga, H., Matsuoka, S.: Hydrogen diffusivity and tensile-ductility loss of solution-treated austenitic stainless steels with external and internal hydrogen. Int. J. Hydrogen Energy **42**(18), 13289–13299 (2017)
33. Díaz, A., Alegre, J., Cuesta, I.: Coupled hydrogen diffusion simulation using a heat transfer analogy **115**, 360–369 (2016)
34. Jemblie, L., Olden, V., Akselsen, O.: A coupled diffusion and cohesive zone modelling approach for numerically assessing hydrogen embrittlement of steel structures. Int. J. Hydrogen Energy **42**(16), 11980–11995 (2017)
35. Charles, Y., Nguyen, H., Gaspérini, M.: FE simulation of the influence of plastic strain on hydrogen distribution during an U-bend test. Int. J. Mech. Sci. **120**, 214–224 (2017)
36. Cui, T., Liu, P., Gu, C.: Finite element analysis of hydrogen diffusion/plasticity coupled behaviors of low-alloy ferritic steel at large strain. Int. J. Hydrogen Energy **42**(31), 20324–20335 (2017)
37. Sezgin, J.-G., Bosch, C., Montouchet, A., Perrin, G., Wolski, K.: Modelling of hydrogen induced pressurization of internal cavities. Int. J. Hydrogen Energy **42**(22), 15403–15414 (2017)

38. Saini, N., Pandey, C., Mahapatra, M.: Effect of diffusible hydrogen content on embrittlement of P92 steel. Int. J. Hydrogen Energy **42**(27), 17328–17338 (2017)
39. Stashchuk, M., Dorosh, M.: Analytical evaluation of hydrogen induced stress in metal. Int. J. Hydrogen Energy **42**(9), 6394–6400 (2017)
40. Polyanskiy, A.M., Polyanskiy, V.A., Yakovlev, Y.A.: Issledovaniye ppotsessov ustalosti i paz-pusheniya metallicheskikh matepialov s ppivlecheniyem metoda oppedeleniya enepgii svyazi vodoroda v tverdom tele (Investigation of the processes of fatigue and destruction of metal by method for determining of the binding energy of hydrogen in solid) **3**, 39–43 (2009)
41. Gorsky, W.S.: Theorie der elastischen Nachwirkung in ungeordneten Mischkristallen (elastische Nachwirkung zweiter Art). Physikalische Zeitschrift der Sowjetunion **8**, 457–471 (1935)
42. Onsager, L.: Reciprocal relations in irreversible processes I. Phys. Rev. **37**(4), 405 (1931)
43. Glansdorff, P., Prigogine, I.: On a general evolution criterion in macroscopic physics. Physica **30**(2), 351–374 (1964)
44. Larché, F., Cahn, J.: The effect of self-stress on diffusion in solids. Acta Metallurgica **30**(10), 1835–1845 (1982)
45. Larché, F. and Voorhees, Peter W.: Diffusion and stresses: basic thermodynamics. In: Beke, D.L. and Szabó, I.A. (eds.) Defect and diffusion forum, vol.129, pp. 31–36. Trans Tech Publ (1996)
46. Drexler, A., Bergmann, C., Manke, G., Kokotin, V., Mraczek, K., Pohl, M., Ecker, W.: On the local evaluation of the hydrogen susceptibility of cold-formed and heat treated advanced high strength steel (AHSS) sheets. Mater. Sci. Eng.: A **140276** (2020)
47. Wu, C.H.: The role of Eshelby stress in composition-generated and stress-assisted diffusion. J. Mech. Phys. Solids **49**(8), 1771–1794 (2001)
48. Maugin, G.A.: On the thermomechanics of continuous media with diffusion and/or weak non-locality. Archive Appl. Mech. **75**(10–12), 723 (2006)
49. Rhode, M., Mente, T., Steppan, E., Steger, J. and Kannengiesser, T.: Hydrogen trapping in T24 Cr-Mo-V steel weld joints–microstructure effect vs. experimental influence on activation energy for diffusion. Welding in the World **62**(2), 277–287 (2018)

Chapter 10
Effect of Hydrogen Concentration and Strain Rate on Hydrogen Embrittlement of Ultra-Fine-Grained Low-Carbon Steel

Evgeniy D. Merson, Pavel N. Myagkikh, Gennadiy V. Klevtsov, Dmitri L. Merson, and Alexei Vinogradov

Abstract During the last few decades, keen attention has been paid to the advanced steels with the ultra-fine-grained (UFG) microstructure manufactured by severe plastic deformation (SPD) techniques. Although these materials often demonstrate prominent mechanical properties, the detrimental environmentally induced effects, such as hydrogen embrittlement (HE), which may appear during their service life, have been just scarcely studied. In particular, the influence of the hydrogen concentration and strain rate, which are among the main factors controlling HE, in general, has not been considered in UFG ferritic steels as yet. Thus, the objective of the present study was to examine the effect of these factors on the mechanical behaviour and fracture mode of the low-alloy steel processed by ECAP in comparison with the conventionally fabricated counterparts. The ECAPed and as-received specimens of the low-alloy steel grade 09G2S were cathodically hydrogen charged at different current densities and then subjected to tensile testing at two different strain rates. The diffusible hydrogen concentration in the specimens before tensile testing was assessed by the hot extraction method. After hydrogen charging both as-received and ECAPed specimens demonstrate HE the extent of which increases with the increasing hydrogen concentration and decreasing strain rate. It is found that the ECAPed steel occludes much higher hydrogen concentration than the as-

E. D. Merson (✉) · P. N. Myagkikh · G. V. Klevtsov · D. L. Merson
Institute of Advanced Technologies, Togliatti State University, Belorusskaya str. 14, Togliatti 445667, Russia
e-mail: Mersoned@gmail.com

P. N. Myagkikh
e-mail: feanorhao@gmail.com

G. V. Klevtsov
e-mail: klevtsov11948@mail.ru

D. L. Merson
e-mail: d.merson@tltsu.ru

A. Vinogradov
Department of Mechanical and Industrial Engineering, Norwegian University of Science and Technology—NTNU, 7491 Trondheim, Norway
e-mail: alexei.vino@gmail.com

V. A. Polyanskiy and A. K. Belyaev (eds.), *Advances in Hydrogen Embrittlement Study*, Advanced Structured Materials 143, https://doi.org/10.1007/978-3-030-66948-5_10

received one. At the given hydrogen concentration, the ECAPed specimens demonstrate stronger hydrogen-induced ductility loss as well as a fundamentally different fracture mode in comparison to the as-received counterparts.

Keywords Hydrogen embrittlement · Equal-channel angular pressing
Low-carbon steel

10.1 Introduction

Grain refinement is one of the most effective methods for strength enhancement in metals, which can be achieved without significant compromising ductility and fracture toughness [1, 2]. Severe plastic deformation (SPD) techniques, among which are equal-channel angular pressing (ECAP), high-pressure torsion (HPT) and many others, have been devised to provide extreme grain size reduction down to the submicrometre scale in many bulk metallic materials including Al [3], Ti [4], Mg [5], Fe-based alloys [6–8], etc. The ultra-fine-grained (UFG) materials obtained in this way demonstrate tremendous strength and ductility balance due to which they received keen attention from material scientists and engineers during the past few decades [9, 10]. In particular, substantial efforts have been made to use the SPD methods for improvement of mechanical properties of low-carbon and low-alloy steels [7, 11, 12]. While the strength of these materials processed by conventional heat treatment is limited to 600–650 MPa, the SPD techniques can provide the increase of the ultimate tensile strength of low-carbon iron up to 1500 MPa which is a very remarkable result [12]. Besides the mechanical properties, the critical factor determining the service performance of structural steels is their resistance to environmentally induced effects such as corrosion, stress corrosion cracking and hydrogen embrittlement (HE). The latter is a well-known phenomenon referred to the deterioration of ductility and other mechanical properties of metals caused by hydrogen, which dissolves and diffuses easily in metals during service life. Fundamental mechanisms of HE are considered and discussed elsewhere in great details [13, 14]. The high-strength steels demonstrate a much higher susceptibility to HE than the mild steels. Therefore, the effect of hydrogen on UFG steels is of substantial practical concern. Although some limited data on HE of the UFG steels can be found in the literature [8, 12, 15, 16], many important issues related to this phenomenon have been not considered as yet. For example, the effect of hydrogen concentration and strain rate, which are best known among the main factors controlling the HE extent, has not been quantified and reported for UFG ferritic steels. Thus, the main objective of the present study was to examine the effect of hydrogen concentration and strain rate on HE of the low-alloy steel with the UFG microstructure processed by ECAP.

10.2 Experimental

The hot-rolled bars of commercial low-alloy steel grade 09G2S with the chemical composition represented in Table 10.1 were subjected to the thermo-mechanical treatment including the following steps: (i) homogenising annealing at 810 °C followed by quenching in water, (ii) tempering at 450 °C, (iii) cold severe plastic deformation by the "CONFORM" ECAP (equal channel angular pressing) [17] process to four passes by the Bc route and (iv) annealing at 350 °C for residual stress removing. Using the electric discharge machine, the as-received and ECAPed bars were longitudinally sliced into 2 mm thick plates from which the specimens for tensile testing with the $4 \times 15\,mm^2$ gauge part, as well as small rectangular specimens for gas analysis with $4 \times 20\,mm^2$ dimensions, were cut out. The surface of all samples was grounded with emery paper to the #240 grade.

The specimens were cathodically hydrogen charged in the 5% H_2SO_4 + 1.5 g/l thiourea solution with the platinum wire used as an anode. To vary the hydrogen concentration, the current density was ranged from 0.3 to 400 mA/cm^2 while the charging time of 1 h was the same in all tests. Within 5 min after hydrogen charging, the specimens were washed with running water and CCl_4 and then subjected to the gas analysis or mechanical testing.

Mechanical testing has been conducted using the screw-driven N50KT (Tinius Olsen) testing machine. Two different nominal strain rates of 5×10^{-3} and $5 \times 10^{-2}\ s^{-1}$ were used (the corresponding traverse velocities were of 5 mm/min and 50 mm/min). After the tensile test, one half of the specimen was subjected to the gas analysis to assess the residual hydrogen concentration, while the other half was used for fractographic observations.

The gas analysis was performed by the hot extraction method using the Galileo G8 (Bruker) gas analyser. To analyse the diffusible hydrogen concentration in steel, each specimen was inserted in the quarts tube of the gas analyser where it was heated up to 200 °C in N_2 gas flux with the heating rate of 17 °C/min and held at the destination temperature for 15 min.

The fractographic and metallographic examinations have been conducted using the scanning electron microscope (SEM) SIGMA (ZEISS) equipped with the electron backscattering diffraction (EBSD) facilities.

Table 10.1 Chemical composition of the steel grade 09G2S, w. %

C	Si	Mn	P	S	Cr	Ni	Cu	V	Al	Fe
0.091	0.722	1.358	0.016	0.011	0.097	0.072	0.235	<0.005	0.018	Base

10.3 Results and Discussion

10.3.1 Microstructure

In the as-received state, the steel has a typical ferrite–pearlite microstructure represented by equiaxed coarse grains of 10 μm average diameter, Fig. 10.1a. The wide assortment of non-metallic inclusions such as MnS is also observed in the microstructure, see the inset in Fig. 10.1a. As can be seen on SEM images presented in Fig. 10.1b, the ECAP results in distortion of the grain boundaries as well as in the stretching of the initial grains in the shear direction at 45° to the extrusion direction. The stretching is more severe at the peripheral part of the ECAPed bar. When examined by the EBSD technique, the microstructure of the ECAPed steel reveals small grains and sub-grains separated by dislocation boundaries, see Fig. 10.1c. The average grain size of the ECAPed steel assessed by EBSD for the grains separated by high-angle boundaries with misorientation angles exceeding 15° is 0.6 μm.

After hydrogen charging, both kinds of specimens exhibit hydrogen-induced defects in the microstructure and on the surface, c.f. Figure 10.1d–f. Hydrogen-induced cracks (HICs) and blisters (HIBs) are present in the microstructure and on the surface of the as-received specimens, respectively, Fig. 10.1d, e. HICs are primarily transgranular, Fig. 10.1d, although some intergranular ones are also can be found, Fig. 10.1e. The number and size of HICs and HIBs in the as-received steel grow with the increasing current density of hydrogen charging. The ECAPed specimens surface reveals no HIBs, yet the HICs, which occasionally connected with the specimen's surface, are readily observed in the microstructure, see Fig. 10.1f. Most of the HICs of the ECAPed steel have the same orientation, which coincides with the grains stretching direction caused by the ECAP simple shear process. These HICs are transgranular with respect to the initial ferritic grains. However, they likely can propagate along with the dislocation sub-grains boundaries.

10.3.2 Gas Analysis

The gas analysis showed that diffusible hydrogen is absent in the uncharged specimens of both kinds. As can be seen in the histogram in Fig. 10.2, the concentration of diffusible hydrogen in the charged specimens increases with the increasing current density. At given charging conditions, the ECAPed steel occludes much higher hydrogen concentration than its as-received counterpart, that is obviously due to the substantially higher density of dislocations and grain boundaries acting as hydrogen traps. One can notice that the significant increase in the hydrogen concentration in the ECAPed steels occurs at the current densities below 20 mA/cm^2, while the further growth of current density up to 400 mA/cm^2 does not affect the hydrogen content considerably. In contrast, the hydrogen concentration in the as-received steel grows significantly as the current density increases from 20 to 400 mA/cm^2. The effect of

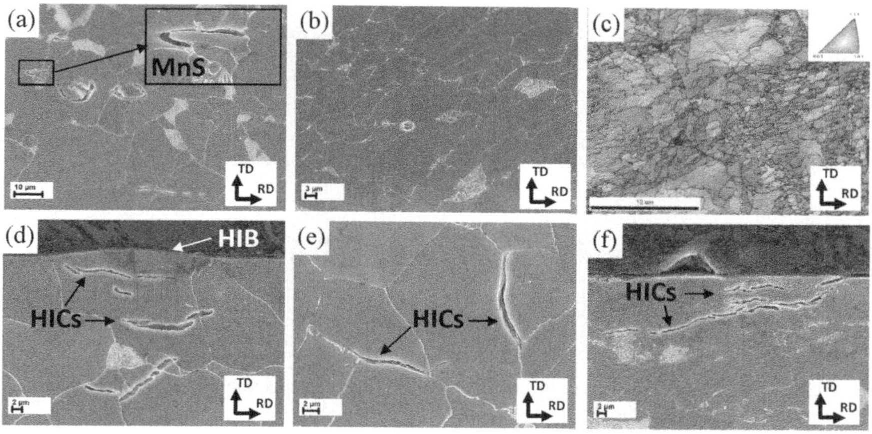

Fig. 10.1 SEM images (**a**, **b**, **d**–**f**) and inverse pole figure (IPF) colored orientation map obtained by EBSD from the microstructure of the as-received (**a**, **d**, **e**) and the ECAPed (**b**, **c**, **f**) specimens of the low-alloy steel grade 09G2S before (**a**–**c**) and after (**d**–**f**) cathodic hydrogen charging

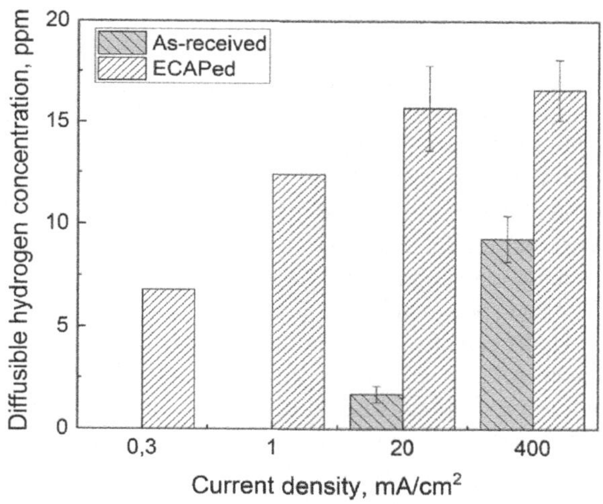

Fig. 10.2 Effect of cathodic hydrogen charging current density on the diffusible hydrogen concentration in the as-received and ECAPed low-alloy steel 09G2S

current density on the hydrogen concentration and hydrogen-induced defects in the same materials has been considered in the dedicated paper in detail [16]. It was shown in particular that the increase of the hydrogen concentration in the as-received steel in the current density range of 20–400 mA/cm^2 is associated with intensive blistering that does not occur in the UFG steel obtained by ECAP.

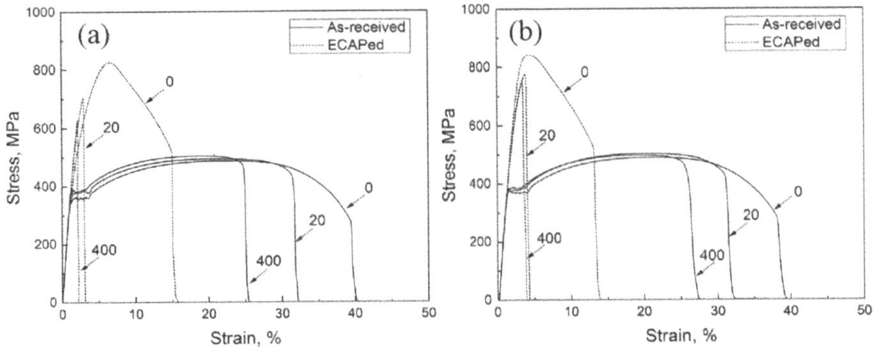

Fig. 10.3 Stress–strain diagrams of the as-received and ECAPed specimens before and after hydrogen charging at low (**a**) and high (**b**) strain rate

10.3.3 Mechanical Testing

The results of the tensile tests showed that the hydrogen-free as-received specimens demonstrate significant elongation to failure and moderate strength. The pronounced yield plateau, strain hardening and necking regions are evident on the stress–strain diagrams of these specimens at low and high strain rates as can be seen in Fig. 10.3. After ECAP, the ultimate tensile strength (UTS) of the steel increases from 485 up to 840 MPa, while the elongation compromises from 40 down to 15%. Besides, the yield plateau vanishes and the strain hardening region becomes hardly visible on the stress–strain diagrams. The increase of the strain rate expectedly provides a slight increase in strength and the decrease in elongation to failure of both as-received and ECAPed specimens.

Hydrogen charging results in substantial ductility drop for the specimens of both kinds. As can be seen in Fig. 10.4a, b, the hydrogen-induced ductility loss of the as-received and ECAPed specimens increases with the increasing hydrogen concentration and decreasing strain rate. At all conditions, the ECAPed specimens show the much more significant ductility loss due to HE than their as-received counterparts. Regardless of the hydrogen concentration and strain rate, the hydrogen charged conventional specimens always fracture after notable yielding, strain-hardening and necking while the ECAPed specimens brake at the stress below the yield point, Fig. 10.3. That is why the hydrogen charging results in significant drop of UTS in the ECAPed steel, while the UTS of the as-received specimens even slightly increases as the hydrogen concentration increases, c.f. Figure 10.4c. For the same reason, the effect of hydrogen concentration on HE of the ECAPed steel is better evaluated by the change in UTS than by elongation or associated ductility loss, which are more informative about the severity of HE in the as-received steel.

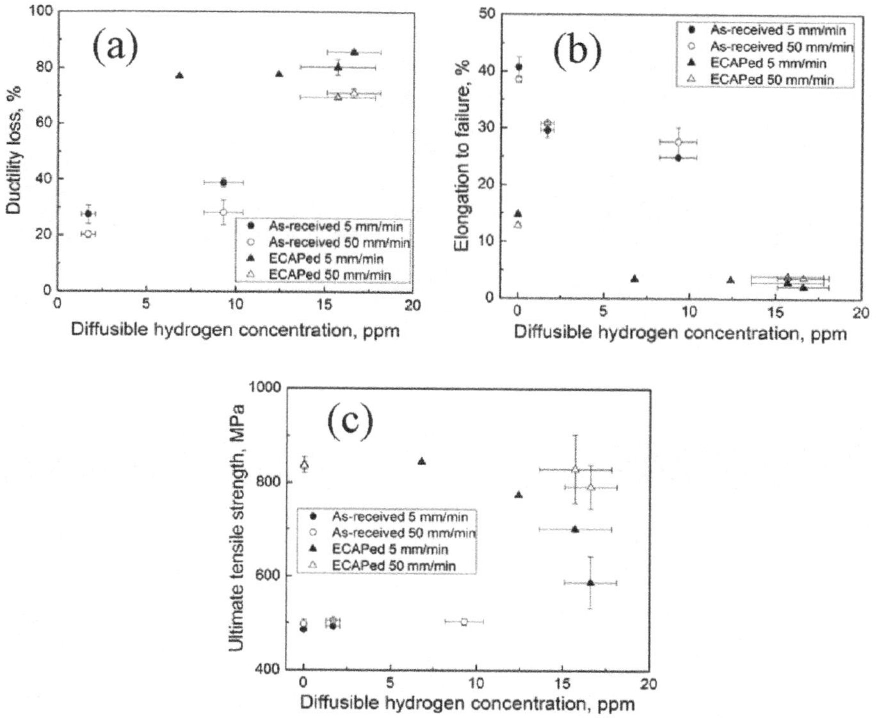

Fig. 10.4 Effect of the diffusible hydrogen concentration on the ductility loss (**a**), elongation to failure (**b**) and ultimate tensile strength (**c**) of the as-received and ECAPed specimens at low and high strain rate of tensile testing

10.3.4 Fractography

The hydrogen-free specimens in the as-received and ECAPed states after tensile testing have completely ductile fracture surface with dimples nucleated at non-metallic inclusions and pearlitic grains, Figs. 10.5a, b, 10.6a, b and 10.7a, b. The fracture surfaces of the hydrogen charged as-received specimens contain "fisheyes"—the round-shape areas with quasi-cleavage morphology and non-metallic inclusions in their centres, Figs. 10.5c–f and 10.7c, e. Depending on the current density and strain rate, the total area of the fisheyes is about 40–50% of the fracture surface area, while the rest is represented by the ductile dimpled relief. Fisheyes are produced by the hydrogen-assisted growth of radial cracks nucleating at non-metallic inclusions. Details of the formation mechanism of these defects and the nature of the quasi-cleavage morphology have been considered elsewhere [8, 18–21].

The quantitative fractographic analysis showed that the increase of both the hydrogen concentration and the strain rate results in the growth of the number of fisheyes as well as in the reduction of their mean area and diameter, Figs. 10.5c–f and 10.8.

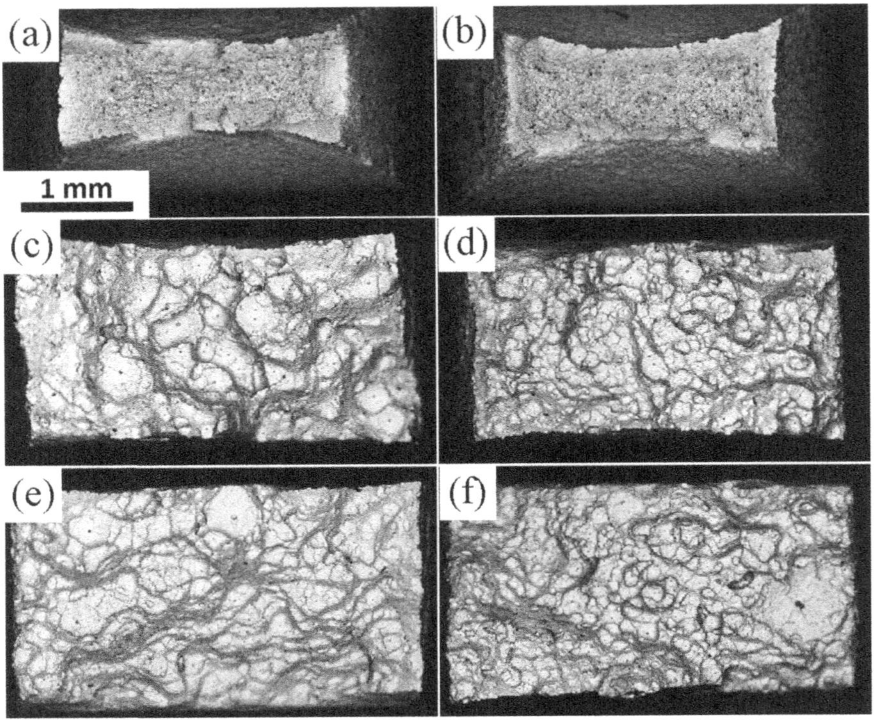

Fig. 10.5 Full-scale views of the fracture surfaces of the as-received specimens of the low-alloy steel grade 09G2S tested at 5 (**a**, **c**, **e**) and 50 mm/min (**b**, **d**, **f**) traverse velocity in the hydrogen-free state (**a**, **b**) and after hydrogen charging at 20 (**c**, **d**) and 400 mA/cm^2 (**e**, **f**). Backscattering electron contrast images obtained by SEM

It is known that the presence of hydrogen at stress risers such as non-metallic inclusions or HICs in steel is a prerequisite for the fisheyes formation [20]. As was shown above, hydrogen charging at the higher current density provides both the increased hydrogen concentration and the greater number of HICs. Since the HICs at the same time act as the points of high hydrogen and stress concentration, and, thus, as the favourable nucleation sights for the fisheyes cracks, the increase of the HICs number should cause the rise of the number of fisheyes. It is reasonable to suppose that during the tensile test, the fisheyes nucleate more readily at the sights with the highest stress and hydrogen concentration. Initiation and propagation of the fisheye crack provide the increase of triaxial stresses ahead of its tip. The regions with high triaxial stresses attract hydrogen, which can diffuse from the neighbouring areas. On the one hand, the supply of extra hydrogen to the crack tip maintains the propagation of this crack and, hence, the further increase of triaxial stress which attracts more hydrogen. On the other hand, the consumption of hydrogen in the neighbouring areas restricts the nucleation of new fisheyes in those areas. However, when the strain rate is high, there

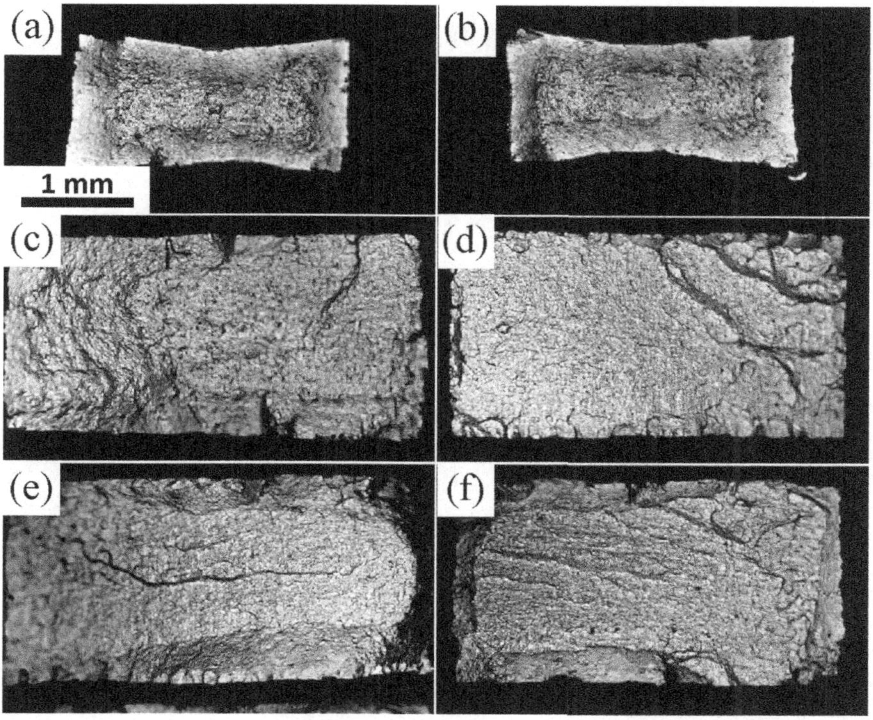

Fig. 10.6 Full-scale views of the fracture surfaces of the ECAPed specimens of the low-alloy steel grade 09G2S tested at 5 (**a**, **c**, **e**) and 50 mm/min (**b**, **d**, **f**) traverse velocity in the hydrogen-free state (**a**, **b**) and after hydrogen charging at 20 (**c**, **d**) and 400 mA/cm^2 (**e**, **f**). Backscattering electron contrast images obtained by SEM

is less time for hydrogen diffusion, while the stress increases quickly so that more fisheyes have a chance to initiate. The aforesaid provides a plausible explanation for the increase in the number of fisheyes and the decrease of their mean dimensions at the increasing strain rate as is observed in the present study.

Regardless of the hydrogen concentration and strain rate, the fracture surfaces of the hydrogen-charged ECAPed steel are entirely brittle, Fig. 10.6c–f. In this case, the fracture surface is seen as a mixture of true cleavage, Fig. 10.7d, and specific tearing morphology, Fig. 10.7f, which is believed to be produced by the hydrogen-assisted cracking along dislocation boundaries of the UFG microstructure of the ECAPed steel [8]. Nevertheless, the exact formation mechanism of the tearing morphology is to be revealed.

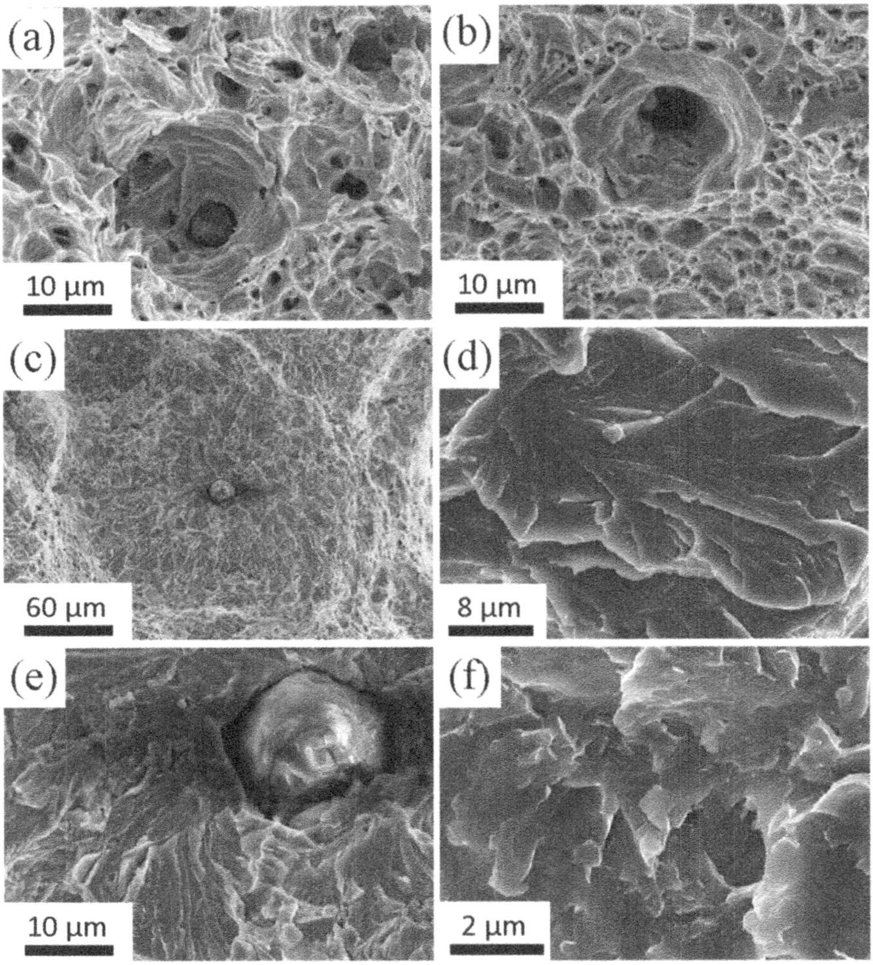

Fig. 10.7 SEM images demonstrating characteristic features of the fracture surfaces of the as-received (**a**, **c**, **e**) and ECAPed (**b**, **d**, **f**) specimens tested before (a, b) and after (**c**–**f**) hydrogen charging: (**a**, **b**) dimpled relief, (**c**) fisheye, (**d**) true cleavage, (**e**) fisheye's quasi-cleavage, (**f**) tearing morphology

10.4 Summary and Conclusions

It is shown in the present study that the microstructure refinement by the ECAP process provides substantial strengthening of the low-alloy steel grade 09G2S, whereas the ductility and the susceptibility to HE are significantly compromised. Moreover, ECAP fundamentally changes the fracture mode of hydrogen-assisted cracking and strongly affects the HE features in the low-alloy steel. The main findings of the present study are as follows.

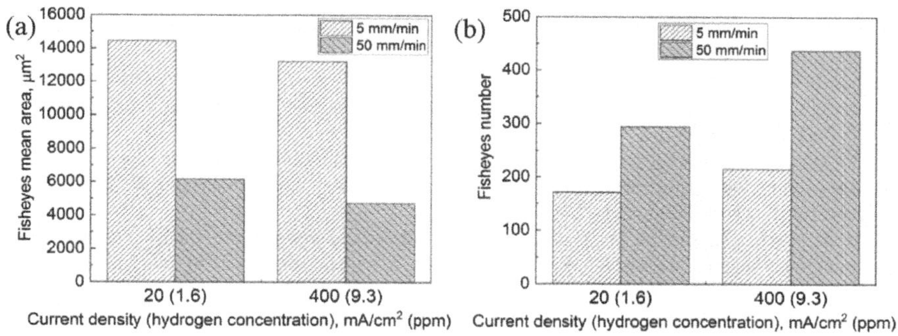

Fig. 10.8 Effect of cathodic hydrogen charging current density (diffusible hydrogen concentration) and strain rate on the mean area (**a**) and number (**b**) of fisheyes in the fracture surfaces of the hydrogen charged as-received specimens of the low-alloy steel grade 09G2S

1. At given cathodic hydrogen charging conditions, the ECAPed steel occludes much higher hydrogen concentration than its as-received hot-rolled counterpart that is apparently due to the increased dislocation density and the volume fraction of grain boundaries serving as hydrogen traps.
2. At given hydrogen concentration and mechanical testing conditions, the ECAPed steel demonstrates more severe ductility loss than its conventional counterpart.
3. The hydrogen charged UFG low-carbon steel obtained by ECAP exhibits HE features which are inherent for most of the high-strength steels embrittled by hydrogen. Specifically, (i) it fractures in the quasi-elastic strain region of the stress–strain diagram, (ii) it shows an entirely brittle fracture surface, (iii) the HE is enhanced at the increasing hydrogen concentration and decreasing strain rate (the latter is also common for other hydrogen embrittled steels and alloys).
4. The fracture surface of the hydrogen charged as-received low-alloy steel is composed of ductile dimpled relief and round-shaped quasi-cleavage regions—fisheyes, which number increases but mean area decreases with increasing hydrogen concentration and strain rate. The fracture surface of the hydrogen embrittled ECAPed steel does not contain fisheyes. However, it is wholly composed of true cleavage and tearing fracture morphology, which formation mechanism is still unknown and is to be revealed.

Acknowledgements The authors appreciate the help of the research team of Prof. R. Z: Valiev (Ufa State Aviation Technical University) for providing the ECAPed material.

References

1. Valiev, R.: Nanostructuring of metals by severe plastic deformation for advanced properties. Nat. Mater. **3**, 511–516 (2004). https://doi.org/10.1038/nmat1180
2. Azushima, A., et al.: Severe plastic deformation (SPD) processes for metals. CIRP Ann. Manuf. Technol. **57**, 716–735 (2008). https://doi.org/10.1016/j.cirp.2008.09.005

3. Khafizova, E., et al.: Microstructure, strength and fatigue of an ultrafine-grained Al-Cu-Mg alloy. Mater. Phys. Mech. **24**, 232–241 (2015)
4. Sabirov, I., et al.: Effect of equal channel angular pressing on the fracture behavior of commercially pure titanium. Metall. Mater. Trans. A **41**, 727–733 (2010). https://doi.org/10.1007/s11661-009-0111-z
5. Vinogradov, A., Serebryany, V.N., Dobatkin, S.V.: Tailoring microstructure and properties of fine grained magnesium alloys by severe plastic deformation. Adv. Eng. Mater. **20**, 1–22 (2018). https://doi.org/10.1002/adem.201700785
6. Hohenwarter, A., Pippan, R.: Fracture of ECAP-deformed iron and the role of extrinsic toughening mechanisms. Acta Mater. **61**, 2973–2983 (2013). https://doi.org/10.1016/j.actamat.2013.01.057
7. Dobatkin, S., Odessky, P.D., Shagalina, S.V.: Ultrafine grained low carbon steels processed by severe plastic deformation. Mater. Sci. Forum. **584–586**, 623–30 (2008). https://doi.org/10.4028/www.scientific.net/MSF.584-586.623
8. Merson, E.D., et al.: Effect of fracture mode on acoustic emission behavior in the hydrogen embrittled low-alloy steel. Eng. Fract. Mech. **210**, 342–357 (2019). https://doi.org/10.1016/j.engfracmech.2018.05.026
9. Estrin, Y., Vinogradov, A.: Extreme grain refinement by severe plastic deformation: a wealth of challenging science. Acta Mater. **61**, 782–817 (2013). https://doi.org/10.1016/j.actamat.2012.10.038
10. Vinogradov, A.: Mechanical properties of ultrafine-grained metals: new challenges and perspectives. Adv. Eng. Mater. **17**, 1710–1722 (2015). https://doi.org/10.1002/adem.201500177
11. Soleimani, F., Kazeminezhad, M.: Synergistic strengthening by severe plastic deformation and post-heat treatment of a low-carbon steel. Steel Res. Int. **89**, 1–10 (2018). https://doi.org/10.1002/srin.201700548
12. Mine, Y., Matsumoto, S., Horita, Z.: Strengthening and hydrogen embrittlement of ultrafine-grained Fe–0.01mass% C alloy processed by high-pressure torsion. Corros. Sci. **53**, 2969–2977 (2011). https://doi.org/10.1016/j.corsci.2011.05.052
13. Robertson, I.M., et al.: Hydrogen embrittlement understood. Metall. Mater. Trans. A **46**, 2323–2341 (2015). https://doi.org/10.1007/s11661-015-2836-1
14. Lynch, S.P.: Hydrogen embrittlement phenomena and mechanisms. Corros. Rev. **30**, 63–133 (2012). https://doi.org/10.1515/corrrev-2012-0502
15. Astafurova, E.G., et al.: hydrogen embrittlement of austenitic stainless steels with ultrafine-grained structures of different morphologies. Phys. Mesomech. **22**, 313–326 (2019). https://doi.org/10.1134/s1029959919040076
16. Merson, E.D., et al.: Effect of equal-channel angular pressing (ECAP) and current density of cathodic hydrogen charging on hydrogen trapping in the low-alloy steel. Lett. Mater. **10**, 152–157 (2020)
17. Raab, G.I. et al.: Long-length ultrafine-grained titanium rods produced by ECAP- conform. Mater. Sci. Forum. **584-586PA**, 80–85 (2008). https://doi.org/10.4028/www.scientific.net/msf.584-586.80
18. Merson, E.D., et al.: Quasi-cleavage hydrogen-assisted cracking path investigation by fractographic and side surface observations. Eng. Fract. Mech. **214**, 177–193 (2019). https://doi.org/10.1016/j.engfracmech.2019.04.042
19. Merson, E.D., et al.: About the nature of quasi-cleavage in low-carbon steel embrittled with hydrogen. Met. Sci. Heat Treat. **61**, 191–195 (2019). https://doi.org/10.1007/s11041-019-00399-x
20. Merson, E., et al.: Quantitative characterization of cleavage and hydrogen-assisted quasi-cleavage fracture surfaces with the use of confocal laser scanning microscopy. Mater. Sci. Eng. A **665**, 35–46 (2016). https://doi.org/10.1016/j.msea.2016.04.023
21. Merson, E., et al.: Application of acoustic emission method for investigation of hydrogen embrittlement mechanism in the low-carbon steel. J. Alloys Compd. **645**, S460–S463 (2015). https://doi.org/10.1016/j.jallcom.2014.12.083

Chapter 11
Wave Nature of Hydrogen Concentration Dynamics in Materials

Alexey V. Porubov, Alexander K. Belyaev, and Vladimir A. Polyanskiy

Abstract The models of the wave description of the hydrogen concentration in materials are discussed and compared. Special attention is paid to the role of nonlinear effects.

Keywords Hydrogen concentration · Bi-continuum · Nonlinear effects
Nonlinear wave modeling

11.1 Introduction

The influence of hydrogen on the plasticity and strength of metals attracts considerable attention in the last years. One can say that the main problem today is hydrogen-induced destruction. The modeling of hydrogen dynamics concerns various approaches. One of them considers the evolution of the hydrogen concentration as a wave process.

The wave description was less developed recently because of the linear description. The hydrogen transfer has been modeled by the equations of diffusion type. However, the linear diffuison equation doesn't possess wave solutions with finite velocity. Only recently, such models began to appear [1–3]. Later, the wave character of concentration has been detected in experiments [3–5]. The model developed in [2] has been extended and thoroughly examined in [5–7]. Also, an influence of the strains in the material on the concentration dynamics has been modeled in [8] on the basis of discrete-continuum nonlinear modeling. All these studies allowed to reveal new phenomena described with the help of a nonlinear description.

A. V. Porubov (✉) · A. K. Belyaev · V. A. Polyanskiy
Institute for Problems in Mechanical Engineering, Bolshoy 61, V.O., Saint-Petersburg, Russia
e-mail: alexey.porubov@gmail.com

Peter the Great St. Petersburg Polytechnic University (SPbPU), Polytechnicheskaya st., 29, Saint Petersburg, Russia

V. A. Polyanskiy and A. K. Belyaev (eds.), *Advances in Hydrogen Embrittlement Study*, Advanced Structured Materials 143, https://doi.org/10.1007/978-3-030-66948-5_11

The aim of the paper is to call attention to the nonlinear wave modeling of the hydrogen concentration and the sources of nonlinearity causing qualitatively new features of the concentration dynamics.

11.2 Bi-Continuum Nonlinear Model

This model has been developed in [2, 3, 6, 7].

Assume the diffuse hydrogen is characterized by m_H, v_H^- and ρ_H^- which are the mass, velocity and volume density respectively. Similarly, the bound hydrogen is described by v_H^+, and ρ_H^+. Then the concentrations are $N_H^- = \rho_H^-/m_H$, $N_H^+ = \rho_H^+/m_H$.

Then basic equations in the one-dimensional case are

$$(\rho_0 + m_H\, N_H^+)\, v_{H,t}^+ + (F_0 - \gamma N_H^+) m_H\, N_H^-(v_H^- - v_H^+) + (\alpha\, N_H^- - \beta\, N_H^+)\, v_H^+ = \sigma_x, \tag{11.1}$$

$$m_H\, N_H^-\, v_{H,t}^- + (F_0 - \gamma N_H^+) m_H\, N_H^- v_H^- + \frac{3}{2} k\, T\, N_{H,x}^- + (\alpha\, N_H^- - \beta\, N_H^+)\, v_H^- = 0, \tag{11.2}$$

$$\rho_{0,t} + (\rho_0\, v_H^+) = 0, \tag{11.3}$$

$$N_{H,t}^+ + (v_H^+\, N_H^+)_x - \frac{1}{m_H}(\alpha\, N_H^- - \beta\, N_H^+) = 0, \tag{11.4}$$

$$N_{H,t}^- + (v_H^-\, N_H^-)_x + \frac{1}{m_H}(\alpha\, N_H^- - \beta\, N_H^+) = 0, \tag{11.5}$$

where k is Boltzmann's constant, T is the absolute temperature of the moving medium, the stress σ is defined in [3].

It was shown in [7] that the solution of the basic equations reduces to finding the solution to the concentration N_H^+,

$$N_H^+ = N_0 + \tilde{N}_H, \tag{11.6}$$

where $\tilde{N}_H$ is the solution to the equation

$$\tilde{N}_{H,tt} - a_1\, \tilde{N}_{H,xx} + a_2 \tilde{N}_{H,t} - a_3 \tilde{N}_{H,xxt} +$$
$$a_4(\tilde{N}_H^2)_{xx} + a_5(\tilde{N}_H \tilde{N}_{H,xt})_x + a_6\, (\tilde{N}_{H,x} \tilde{N}_{H,t})_x = 0, \tag{11.7}$$

where

$$a_0 = \frac{m_H}{\alpha},\; a_1 = \frac{3kT\beta((F_0 - \gamma N_0)\, m_H + 2\alpha)}{2(F_0\, m_H + \alpha)^2 m_H},\; a_2 = \frac{\alpha + \beta}{m_H},$$

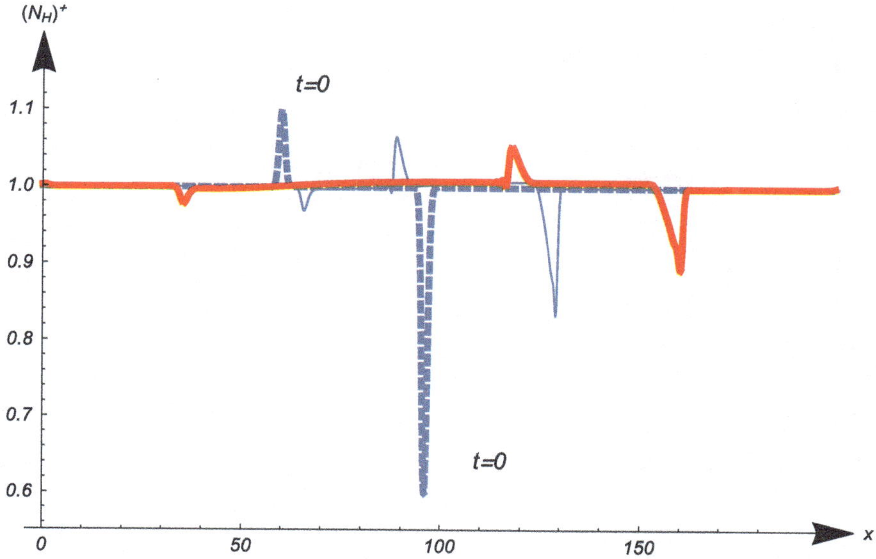

Fig. 11.1 Dynamics of concentration at the coefficients of Eq. (11.7): $a_1 = 1, a_2 = 0.01, a_3 = 0.01, a_4 = 0.75, a_5 = 0.075, a_6 = 0$. The profiles are shown in times 0, 30, 60

$$a_3 = \frac{3kTm_H((F_0 - \gamma N_0)\, m_H + 2\alpha - \beta)}{2(F\, m_H + \alpha)^2 m_H},\ a_4 = \frac{3kT\gamma\beta}{4(F_0\, m_H + \alpha)^2}$$

$$a_5 = \frac{3kTm_H\gamma}{2(F_0\, m_H + \alpha)^2},\ a_6 = \frac{3kT(\alpha - \beta)}{2(F\, m_H + \alpha)^2 N_0}.$$

The physical meaning of the coefficients in Eq. (11.7) relates to an influence of the corresponding terms in the equation on the dynamics of concentration. Thus, the dominance of coefficient a_1 results in a hyperbolic wave nature of the propagation of the concentration front described by the terms with the second-order spatial and temporal derivatives, while the terms with coefficients a_2 and a_3 are responsible for the decay of the concentration wave. The coefficients $a_4 - a_6$ at nonlinear terms in Eq. (11.7) characterize the influence of nonlinearity, see [7] for details.

Numerical simulations demonstrate the influence of different nonlinear terms on the dynamics of localized variation in concentration. Numerical simulations are performed by assuming $N_0 = 1$ and considering evolution of a localized input defined by a Gaussian distribution,

$$\tilde{N}_H = N_0 + Q_1 \exp(-(x - x_0)^2/Q_2) + Q_3 \exp(-(x - x_1)^2/Q_2) \text{ at } t = 0, \tag{11.8}$$

where x_i and Q_i, are constants. A unidirectional evolution is provided by an extra nonzero initial condition for the initial velocity,

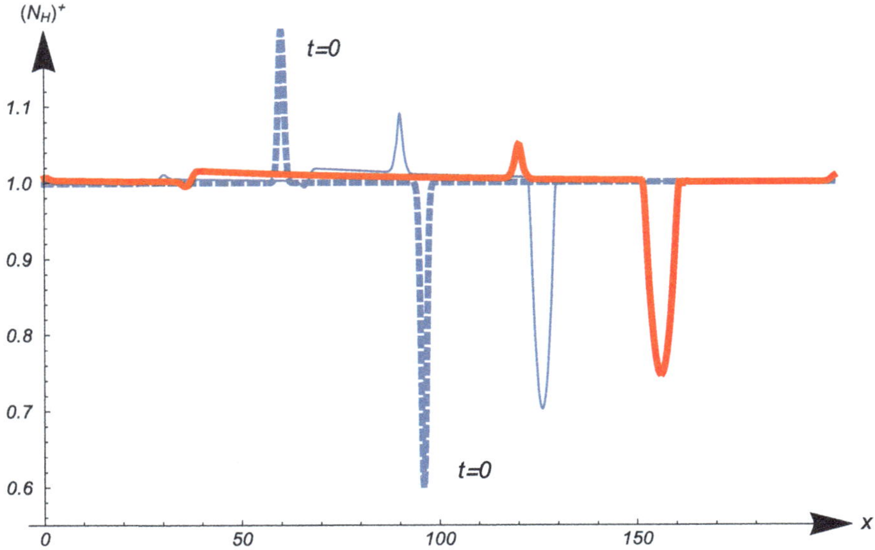

Fig. 11.2 Dynamics of concentration at the coefficients of Eq. (11.7): $a_1 = 1, a_2 = 0.01, a_3 = 0.01, a_4 = 0, a_5 = 0, a_6 = 0.5$. The profiles are shown in times 0, 30, 60

Shown in Fig. 11.1 is the evolution of the initial condition containing both negative and positive parts, Q_1 and Q_3 are of either sign in Eq. (11.8). They are marked by $t = 0$. One can see those initial disturbances propagate losing their amplitudes. Besides, a decrease in the amplitude, one can see a development of an asymmetry of the profile: the negative part suffers a steepness of the front edge of the wave, while the positive part demonstrates a steepness of the back edge of the wave.

On the contrary, an influence of the nonlinear term $(\tilde{N}_{H,x}\tilde{N}_{H,t})_x$ on the same initial condition doesn't provide an asymmetry as shown in Fig. 11.2. There is a decrease in the amplitude for both the positive and negative parts of the input, also one can see an increase in the width of the propagating disturbance of concentration.

More simulations about relative influence of nonlinearities and an influence of the polarity of the input on the behavior of the localized wave of concentration can may be found in Ref. [7].

11.3 Dicrete Continuum Nonlinear Model

The decrease in the amplitude of concentration wave happens due to an influence of dissipative terms in Eq. (11.7). The dynamical behavior of concentration varies when the strains in the material are taken into account.

The crystalline structure of a material can be described by a one-dimensional chain with masses connected by elastic springs and only with neighboring interactions between masses when the interaction obeys the Hookean linear law. Then the equation of motion for a mass with the number n is

$$m\ddot{u}_m = C_0(u_{n+1} - u_n) - C_0(u_n + u_{n-1}), \tag{11.9}$$

where u_i is the displacement of the i-th mass in the chain, m is the mass of the element of the chain, C_0 is the constant stiffness.

We assume a rheological relationship, $C = C_n(n, t)$,

$$\frac{N_0 + N_n}{C_n} = \frac{N_0}{C_0} + \frac{N_n}{C_H}, \tag{11.10}$$

where N_0 is the known constant concentration of the elements of the chain in the material, $N_n(n, t)$ is the discrete concentration of hydrogen, C_H is some known constant, $N_n(n, t)$ is the concentration of hydrogen in nanovoids, and C_H characterizes the weakening of the material due to the formation of nanovoids. Then we obtain

$$C_n = \frac{(N_0 + N_n)C_H C_0}{C_H N_0 + C_0 N_n}. \tag{11.11}$$

The statement of the problem is described more precisely in [8].

The continuum concentration $N(x, t)$ is usually described by an equation of transfer,

$$N_t + (\beta_0 + \beta_1 u_x)N + \tilde{\delta} u_x + \gamma N_{xx} = 0. \tag{11.12}$$

where a diffusion of the hydrogen concentration, N_{xx} is taken into account as well as a contribution of strains in the dynamics of concentration. First, it affects the source term coefficient $\beta_0 + \beta_1 u_x$, second, there is an influence δu_x to the variation of concentration. β_0, β_1, $\tilde{\delta}$, γ are constants.

The weakly long-wavelength case is considered and a small parameter ε is introduced,

$$u = \varepsilon u(X, T, , \tau, Z), y = \varepsilon^2 y(X, T, \tau, Z),$$

where

$$X = \varepsilon x, \ T = \varepsilon t, \ \tau = \varepsilon^3 t, , \ Z = \varepsilon^4 t.$$

where

$$y(x, t) = \frac{C_0}{C_H}\frac{N}{N_0}.$$

It is shown in Ref. [8] that for small y one assumes

$$y = y_0 + \varepsilon y_1 + ...$$

and the governing continuum equation for y_0 is

$$2\,h\beta_0\sqrt{C_0 m}\,y_{0,\theta\tau} + \frac{\beta_0 C_0 h^4}{12} y_{0,\theta\theta\theta\theta} - \frac{\beta_0 C_0 h^2}{2}(y_0^2)_{\theta\theta} = 0, \tag{11.13}$$

where

$$\theta = X - \sqrt{\frac{C_0}{m}}T. \tag{11.14}$$

The traveling wave solution is [8]

$$y_0 = F - h^2\kappa^2 \mathrm{sech}^2(\kappa\xi), \tag{11.15}$$

where

$$\xi_\theta = 1, \xi_\tau = -W, \ W = \frac{h\sqrt{C_0}}{6\sqrt{m}}\left(h^2\kappa^2 - 3F\right)$$

The parameter W should be

$$W < -\frac{\kappa^2 h\sqrt{m}}{3\sqrt{C_0}}.$$

to achieve only positive values of the solution for concentration.

Numerical results shown in Fig. 11.3 are the dynamics of an initial disturbance of concentration around constant value F. The initial profile contains both positive

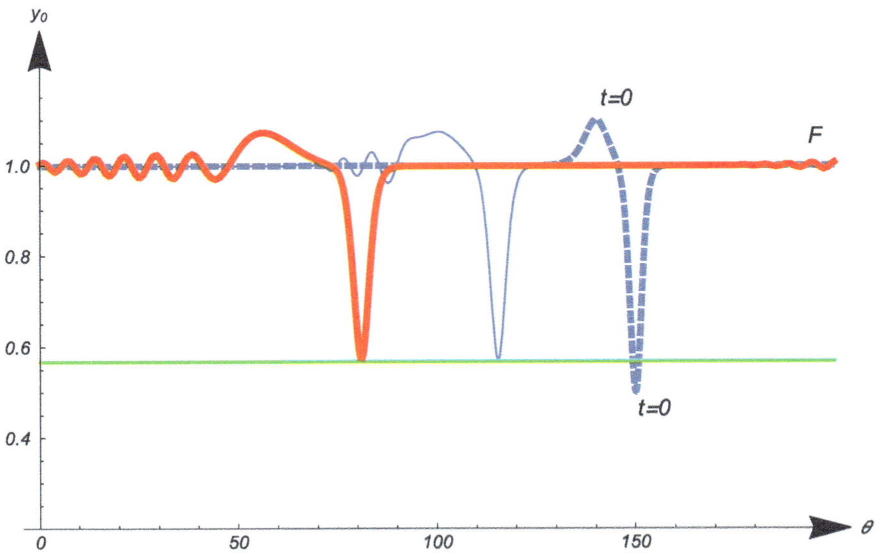

Fig. 11.3 Dynamics of concentration The constant $F = 1$

and negative amplitude disturbances marked by $t = 0$. As times goes on, the negative amplitude part evolves into traveling wave with a constant amplitude which is illustrated by a horizontal line in Fig. 11.3. The wave travels to the left. The positive amplitude disturbance is dispersed and doesn't give rise to a traveling localized wave of concentration.

The results obtained are written in the transformed coordinates. Coming back to the original variables, one obtains the solution for the stiffness [8]

$$C = C_0(1 - \varepsilon^2 F + \varepsilon^2 \kappa^2 h^2 \mathrm{sech}^2(\varepsilon\kappa(x - (\sqrt{\frac{C_0}{m}} + \varepsilon^2 W)t). \tag{11.16}$$

The initial constant concentration F results in a decrease in the stiffness. The propagating wave of concentration locally increases the stiffness but not above C_0. The wave moves in the same direction as those shown in Figs. 11.1 and 11.2. More numerical solutions can be found in Ref. [8].

11.4 Conclusions

Several related phenomena, the formation of nanovoids associated with the hydrogen accumulation, the weakening of the material as a result of the nanovoids formation, and the non-uniform distribution of hydrogen and nanovoids, which is observed during the tension of corset specimens without stress concentrators with a uniform initial hydrogen concentration at a uniform uniaxial deformation, are described using the nonlinear wave propagation.

The dynamics or the relatively high rate of these processes can be explained by the fact that hydrogen itself does not move inside the material, but changes its state passing from a diffuse phase to a gaseous one inside nanovoids which can appear and disappear when the stress and strain change. Most likely, the consequences of this transformation for the strength and internal microstructure of the metal depend on the parameters of the material, loading rate, temperature, and value of the initial hydrogen concentration. The proposed wave approach makes it possible to establish these relations after identifying the parameters of the models.

Two models are discussed. The first one concerns the modeling of a bi-continuum. Numerical simulations reveal different influences of the nonlinear terms appearing in the presence or in absence of the inhomogeneous force. Qualitatively different effects such as arising of the tail behind the localized wave and formation of the counterpart wave are found. Also, the nonlinear discrete-continuum model of mutual influence of the hydrogen concentration and longitudinal strain in a chain is developed. A nonlinearity is introduced via the variations in the coefficient of stiffness of the springs between the elements of the chain. A two-dimensional lattice model will be the subject of our future work to describe both an increase and a decrease in the elastic constants.

Acknowledgements The study of AKB and VAP was funded by the Russian Foundation for Basic Research, projects 20-08-01100 and 18-08-00201.

References

1. Smirnov, L.I.: On the wave propagation of concentration hydrogen disturbances in palladium. Phys. Metals Metallogr. **109**(2), 107–110 (2010)
2. Indeitsev, D., Semenov, B.: About a model of structural-phase transformations under hydrogen influence. Acta Mech. **195**(1–4), 295–304 (2008)
3. Indeitsev, D., Polyanskii, V., Semenov, B., Sterlin, M., Yakovlev, Y.: Hydrogen diffusion in metals under fatigue failure. In: 19th European Conference on Fracture: Fracture Mechanics for Durability, Reliability and Safety, ECF 2012, Kazan, 26–31 August 2012
4. Belyaev, A.K., et al.: Surface effect of the waves of plastic deformation and hydrogen distribution in metals. In: 2017 Days on Diffraction (DD), pp. 45–50. IEEE (2017)
5. Polyanskiy, V.A., Belyaev, A.K., Porubov, A.V., Yakovlev, Y.A., Polyanskiy, A.M.: Nonlinear waves of the internal medium due to dynamic loading of bi-continuum. In: 2019 Days on Diffraction (DD), pp.153–158. IEEE (2019)
6. Belyaev, A.K., Polyanskiy, V.A., Porubov, A.V.: Nonlinear dynamics of hydrogen concentration in high-strength and high-entropy alloys. Continuum Mech. Thermodyn. **31**, 17851794 (2019)
7. Porubov, A.V., Belyaev, A.K., Polyanskiy, V.A.: Nonlinear modeling of dynamics of hydrogen concentration in alloys. Commun. Nonlinear Sci. Numer. Simulat. **90**, 105402 (2020)
8. Porubov, A.V., Belyaev, A.K., Polyanskiy, V.A.: Nonlinear hybrid continuum-discrete dynamic model of influence of hydrogen concentration on strength of materials, Continuum Mech. Thermodyn. (2020). https://doi.org/10.1007/s00161-020-00936-7
9. Zabusky, N.G., Deem, G.S.: Dynamics of nonlinear lattices I. Localized optical excitations, acoustic radiation, and strong nonlinear behavior. J. Comp. Phys. **2**, 126–153 (1967)
10. Andrianov, I.V., Awrejcewicz, J., Weichert, D.: Improved continuous models for discrete media. Math. Probl. Eng. (Open Access) **986242** (2010)

Chapter 12
Characterization of Hydrogen Trapping Systems and HIC Susceptibility of X60 Steel by Traditional and Innovative Methodologies

Renzo Valentini, Francesco Aiello, Linda Bacchi, Fabio Biagini, Serena Corsinovi, and Michele Villa

Abstract Hydrogen interaction with steels and metallic alloys, in general, is an old issue, but the interest in the phenomenon is incredibly increased in recent years. From the '90s of the last century up to today, the papers about Hydrogen Embrittlement (Scopus, Elsevier, Amsterdam, [1]) are almost tripled! This establishes the growing attention on this topic, due also to the development of hydrogen-based economy. In Oil & Gas industry, hydrogen–steel interaction and related phenomena are often related to corrosion reactions, a consequence of the severe sour environments typical of this industrial sector. This is the reason why corrosion is the second most frequent cause of pipeline failures (Yang et al., Reliab Eng Syst Saf 159, 214–222, [2]), with the peculiar result of dangerous substances released in the environment, and it has translated into a continuous development of new technologies to monitor and control the 'corrosion state'. The present work aims at characterizing the hydrogen susceptibility of X60 steel, a High-Strength Low-Alloy (HSLA) steel, widely used in Oil & Gas industry. The study is carried out by means of a rigorous approach based on traditional scientific techniques; moreover, an innovative solution was developed, validated and proposed to approach the possibility of on-field monitoring of operating pipelines.

Keywords Hydrogen embrittlement · Hydrogen-induced cracking · X60 · Pipeline steels · Sour environment · Hydrogen trapping · Hydrogen diffusion

---

R. Valentini (✉) · F. Aiello
Department of Civil and Industrial Engineering, University of Pisa, 56122 Pisa, Italy
e-mail: renzo.valentini@unipi.it

L. Bacchi · F. Biagini · S. Corsinovi · M. Villa
Letomec SRL, 56126 Pisa, Italy

V. A. Polyanskiy and A. K. Belyaev (eds.), *Advances in Hydrogen Embrittlement Study*, Advanced Structured Materials 143, https://doi.org/10.1007/978-3-030-66948-5_12

12.1 Introduction

Hydrogen embrittlement is a phenomenon of great interest for Oil&Gas industry, due to the use of higher grades steels, as well as low-purity service fluids, containing H_2S, for example. In this scenario, the possibility of approaching the plants and pipelines on-field monitoring looks very attractive and could be the real turning point for reducing failures and manage safe planned maintenance.

The mechanisms of hydrogen embrittlement and hydrogen interaction with metal lattice are still under investigation, but some theories are overall accepted to demonstrate the dangerous and deteriorating outcome of hydrogen absorption on steels' mechanical properties and behaviour. Nevertheless, the effect of hydrogen is mainly and deeply investigated for low-carbon steels, stainless steels, TRIP steels, DP steels, pipeline steels and so on [3], it is well known that the embrittlement effect is particularly severe for high-strength and ultra-high-strength steels. Typically, a UTS higher than 1000 MPa is assumed as threshold value to significantly appreciate it.

The most diffused models that describe the hydrogen embrittlement in metals are the HELP (Hydrogen-Enhanced Local Plasticity) and HEDE (Hydrogen-Enhanced DEcohesion). These models are quite different, HELP is related to hydrogen enhancement of dislocations' generation and motion, giving rise to local plasticized zones and material softening. Other literature studies state the opposite phenomenon, that is the pinning of dislocations due to the presence of hydrogen atoms. Katzarov et al. [4] and Taketomi et al. [5] observed and demonstrated the coexistence of both actions. Hydrogen affects dislocation mobility as a function of several factors, including hydrogen concentration, tension stress state and temperature. HEDE model is instead based on the reduction of metal lattice chemical bounds due to the presence of hydrogen, which promotes crack formation and propagation. It is strictly correlated to locally reach the critical hydrogen concentration.

Currently, the most recent researches propose a new approach which consists of the synergistic contribution of HELP and HEDE theories, the so-called HELP-mediated HEDE model [6]. A first HELP stage is followed by a combination of HELP and HEDE mechanisms, where hydrogen enhances the mobility of dislocations (HELP), giving rise to high-density dislocations pile up and plasticization at the crack tip. At the same time, hydrogen transportation by dislocation increases the local concentration in the metal leading to decohesion of chemical bonds (HEDE).

Hydrogen-induced cracking (HIC) is another phenomenon typical of Oil & Gas industry. It usually takes place near to precipitates or non-metallic inclusions, vacancies or other microstructural defects, and consists in crack nucleation and propagation even without applied load to the part or component. When hydrogen accumulates in these areas, it can recombine into molecular hydrogen and reach a local overpressure level able to crack the metal according to hydrogen pressure theory [7].

Peng et al. [7] stated the HIC nucleation depends on the type and shape of the inclusions: MnS, as well as Al, Ti and Ca oxides, are very active initiation sites, while inclusions rich in Si present a low risk. Recent studies [7] showed a better resistance of pipeline steels with controlled precipitation of carbides and nitrides (as V_4C_3, TiC, NbC and nanosized (Ti,Mo,C) aimed at reducing particles size and increasing their number [8].

Sour environment, typical of petrochemical plants, is characterized by the presence of a strongly reactive and dangerous compound, the hydrogen sulphide. H2S inhibits the recombination of atomic hydrogen into molecular hydrogen according to the following reactions:

$$H_2S + e^- \rightarrow H + HS \tag{12.1}$$

$$HS^- + e^- \rightarrow H + S^{2-} \tag{12.2}$$

$$H^+ + e^- \rightarrow H. \tag{12.3}$$

Equation (12.3) allows atomic hydrogen to be adsorbed on the metal surface, considerably increasing the probability of absorption in the bulk material and the risk of hydrogen damage [9].

12.2 Materials and Methods

The present work focuses on API X60 and Iron-Armco§. API X60 is HSLA pipeline steel that combines good resistance, ductility, weldability and corrosion resistance. The strengthening mechanisms consist of the partial precipitation of inclusions, coupled with grain size reduction and it is subjected to hot rolling. Fe Armco is instead cold rolled and annealed. Mechanical properties and chemical compositions are listed in Tables 12.1 and 12.2 respectively.

For permeation tests, rectangular sheet specimens of different thickness (0.5, 1 and 2 mm) were used, for HIC tests and TPD analyses, cylindrical specimens were manufactured, of diameter, respectively, 13 and 6 mm.

12.3 Experimental

A series of experimental tests were carried out in order to characterize the steel behaviour in a hydrogen environment and its susceptibility to hydrogen embrittlement.

Table 12.1 Mechanical properties of API X60

R_{p02} (Mpa)	R_m (Mpa)	Elongation (%)
436	520	40

Table 12.2 Composition of Iron Armco § and API X60

	API X60	Fe Armco
C	0.1343%	0.01%
Mn	1.15%	0.15%
Si	0.2523%	–
S	0.0047%	0.015%
P	0.0217%	0.015%
Al	0.0254%	0.05%
Ti	0.022%	–
Nb	0.038%	–
N	0.0102%	0.007%
Fe	Balance	Balance

12.3.1 Permeation Tests

Permeation tests were performed to evaluate hydrogen diffusion in steels, determining the hydrogen diffusion coefficient.

The traditional test method, based on Devanathan–Stachurski double cell [10], was used. The Devanathan–Stachurski technique involves the application of a Pd coating to the sample surface. A solution of $PdCl_2$ (5 g/L) and NH3 (28% in volume) was used and a current density of 4 mA/cm^2 [11] was applied, after a prior surface polish. The sample is placed between two semi-cells and surface passivation of one side is achieved by using 0.1 N NaOH solution and imposing 200 mV with respect to Ag/AgCl reference electrode.

Permeation tests were also carried out by means of HELIOS 2 (Hydrogen Embrittlement Line Instruments and Operative Sensors), an innovative equipment patented by Letomec S.r.l [12] (patent EP2912452B1), and results were compared.

HELIOS 2 is based on an electrochemical cell, where the metallic sample, the cathode, is in contact with the electrolytic solution through a hole in the cell with calibrated diameter. Applying a certain current between the anode and the cathode, hydrogen is produced on the exposed surface of the sample. A probe, containing a solid-state gas sensor, registers the hydrogen flux coming out from the other side (see Fig. 12.1). HELIOS method since there is no electrolytic solution on the detection side, doesn't require the passivation of the metallic sample.

After the preliminary comparison among Devanathan and HELIOS 2 methods on the reference material (Iron Armco), permeation tests were repeated on API X60 HSLA steel. Tests were executed by varying operative temperature, to estimate the low-energy traps binding energy and, in a second phase, tests on samples with different thicknesses were executed. Finally, for what concerns permeation tests, a real case of cracking was simulated. A sheet of 2 mm thickness has been put in contact with H_2SO_4 1 N and As_2O_3 10 mg/L and a current density of 50 mA/cm^2 was applied.

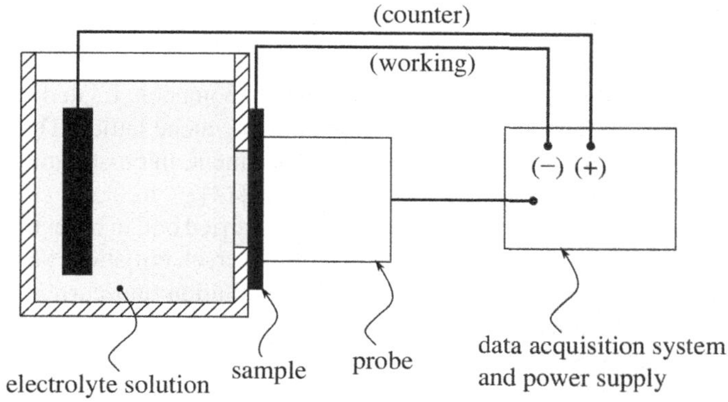

Fig. 12.1 Helios 2 equipment scheme

12.3.2 TPD Tests

Hydrogen trapping systems were studied by the temperature-programmed desorption technique. Pre-charged samples were heated from room temperature up to 600 °C. Various heating rates were applied, and binding energy of traps was extrapolated. According to Choo and Lee model [13], the heating rate can be expressed as follows:

$$\dot{z} = A(1 - z)e^{E_a/RT} \tag{12.4}$$

with $z = (C_0 - C(t))/C_0$ and where C_0 and $C(t)$ represent, respectively, the initial and the instantaneous hydrogen concentration in steel, A is a proportionality constant, R the gas constant, T the absolute temperature and E_a the activation energy of a trap. Deriving the Equation (12.4), the following expression is obtained:

$$\frac{d\dot{z}}{dt} = -A\dot{z}e^{-\frac{E_a}{RT}} + A\,(1 - z)\,e^{-\frac{E_a}{RT}}\,\frac{E_a}{RT^2}\,\frac{dT}{dt} \tag{12.5}$$

imposing $\frac{d\dot{z}}{dt} = 0$, the maximum of the function ($\frac{dT}{dt} = \phi$) can be determined.

$$-A\dot{z}e^{-\frac{E_a}{RT_p}} + \dot{z}\frac{E_a}{RT_p^2}\phi = 0 \overset{(T_p=\text{Peak Temperature})}{\Longrightarrow} -\frac{E_a}{RT_p} = \ln\left[\left(\frac{E_a}{ART_p^2}\right)\phi\right] \tag{12.6}$$

In a diagram where $\left(\frac{1}{T_p};\ \ln\left[\left(\frac{\phi}{T_p^2}\right)\right]\right)$ were plotted, experimental points settle on a straight line and the binding energy can be estimated from the slope of that line.

12.3.3 HIC Tests

HIC is an extremely dangerous and unpredictable phenomenon, related to corrosion phenomena and consequent hydrogen absorption in the metal lattice. There are a lot of factors influencing HIC phenomenon like environment, microstructure, shape of the grain boundaries, number and type of inclusions [14].

That is why, a series of HIC nucleation tests were carried out, in order to determine the critical hydrogen concentration able to give rise to crack initiation. Samples were electrochemically charged by varying electrolytic solution and current density to introduce a different amount of hydrogen in the metal lattice, then they were cut into two parts. A half was observed under light microscope to check the presence of cracks, the other half was used for hydrogen measurement by means of thermal desorption analysis. Figure 12.2 shows an example of cracked material in a hydrogen-induced cracking test.

12.3.4 Numerical Analysis

A series of numerical simulations were also performed to determine hydrogen trapping parameters in the bulk steel. Typical diffusion behaviour, explained by Fick's law [15, 16], is not verified in the presence of hydrogen trapping sites in the metal lattice. McNabb and Forster developed a theory to consider this effect and this was implemented in a simulation model to determine all kinetic parameters [17].

$$\begin{cases} \frac{\partial C}{\partial t} = D\frac{\partial^2 C}{\partial x^2} - N_r\frac{\partial v}{\partial t} - N_i\frac{\partial w}{\partial t} \\ \frac{\partial v}{\partial t} = k_r C(1-v) - pv \\ \frac{\partial w}{\partial t} = k_i Ct(1-w) \end{cases} \tag{12.7}$$

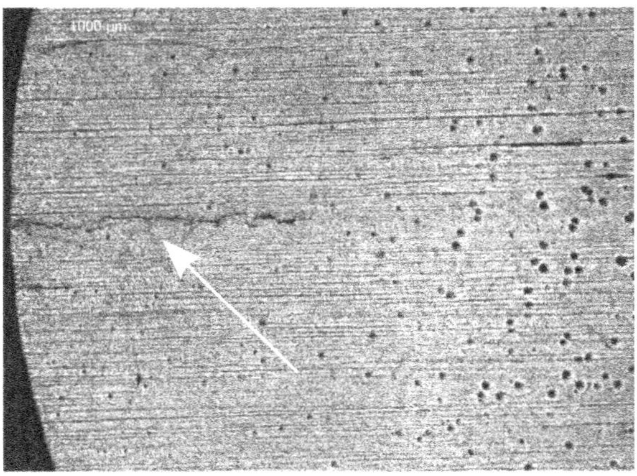

Fig. 12.2 API X60 sample, hydrogen crack was highlighted with the white arrow

where C and D are the hydrogen concentration and the diffusion coefficient, respectively; N_r and N_i are the reversible and irreversible traps concentration, ν is the occupied reversible trap fraction, w the irreversible one; k_r is the reversible trapping rate, k_i the irreversible one and p is the release amount for reversible traps.

Solve Eq. 12.7, combined with the corresponding initial and boundary conditions, allows to determine the outlet hydrogen flux defined as $J(t) = -D\partial_x C$. The initial conditions correspond to reversible and irreversible traps initially empty and material free of diffusible hydrogen. Boundary conditions consist of a constant concentration C_0 on the face of the specimen in contact with the electrolyte solution and zero concentration on the opposite face. The mathematical formulation is reported in Eq. (12.8) and Eq. (12.9), respectively, where a is the sample thickness.

$$C(t = 0, x) = 0; \quad \nu(t = 0, x) = 0; \quad w(t = 0, x) = 0 \tag{12.8}$$

$$C(t, x = 0) = C_0; \quad C(t, x = a) = 0 \tag{12.9}$$

The hydrogen outlet flux is a function of time and seven other parameters C_0, D, k_r, N_r, N_i, k_i, p. The iron lattice diffusion coefficient is constant at room temperature and equivalent to $7 \cdot 10^{-9}$ m^2/s [18]. C_0 can be easily obtained from the hydrogen flow during steady state. The k_r/p ratio can be determined from the knowledge of the lattice number of sites per unit of volume N_L and from the activation energy of traps E_a known by the TPD technique using Eq. (12.10) [19].

$$\frac{k_r}{p} = \frac{1}{N_L} \exp\left(\frac{E_a}{RT}\right) \tag{12.10}$$

The remaining parameters were determined by minimizing the cost function f defined in Eq. 12.8.

$$F = \min[J_{\text{simulated}} - J_{\text{experimental}}]^2 \tag{12.11}$$

12.4 Results

12.4.1 Permeation Tests

Permeation experiments were first executed on Iron Armco. Samples were tested with Devanathan method, with and without Pd coating, besides HELIOS 2 technique on bare specimens. Results were presented in Table 12.3.

Note that Palladium coating on the reading side is widely shown and demonstrated in the international literature to be a fundamental feature to get significant results [11], and this effect is clearly confirmed in Fig. 12.3.

Table 12.3 Permeations test results obtained with different detecting techniques on Fe-Armco

Thickness (m)	Material	J (mol/m s)	D (m^2/s)
0.0005	Fe D	$6.96 \cdot 10^{-14}$	$1.42 \cdot 10^{-10}$
	Fe + Pd D	$1.22 \cdot 10^{-12}$	$1.23 \cdot 10^{-10}$
	Fe H	$1.18 \cdot 10^{-12}$	$7.76 \cdot 10^{-11}$
0.001	Fe D	$3.74 \cdot 10^{-14}$	$2.45 \cdot 10^{-10}$
	Fe + Pd D	$1.35 \cdot 10^{-12}$	$4.60 \cdot 10^{-10}$
	Fe H	$9.12 \cdot 10^{-13}$	$2.22 \cdot 10^{-6}$
0.002	Fe D	$4.38 \cdot 10^{-14}$	$1.35 \cdot 10^{-9}$
	Fe + Pd D	$1.27 \cdot 10^{-12}$	$1.38 \cdot 10^{-9}$
	Fe H	$1.29 \cdot 10^{-12}$	$8.87 \cdot 10^{-10}$

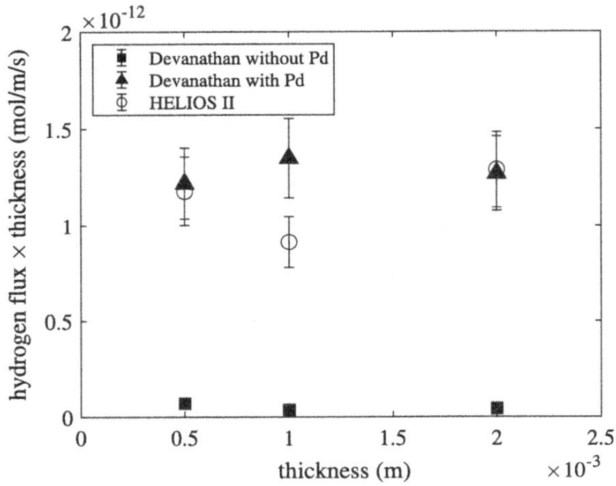

Fig. 12.3 Comparison of exiting hydrogen flux in three different test conditions and for various sample thicknesses

12.4.1.1 API X60 Steel

Helios test results as a function of temperature were shown in Fig. 12.4. From these tests, the binding energy of the traps [20] was extrapolated and resulted equal to 18 kJ/mol.

Figure 12.5 indicated the stationary hydrogen flux determined during experimental results as a function of sample thickness.

Another test performed has been the permeation in crack condition (cracking conditions evaluated from HIC studies): so the conditions in terms of solution and current density able to generate microvoids due to the hydrogen interaction with metal lattice.

The permeation curve through the metal wall of the cracking case was reported in Fig. 12.6.

Fig. 12.4 Permeation tests at different temperatures

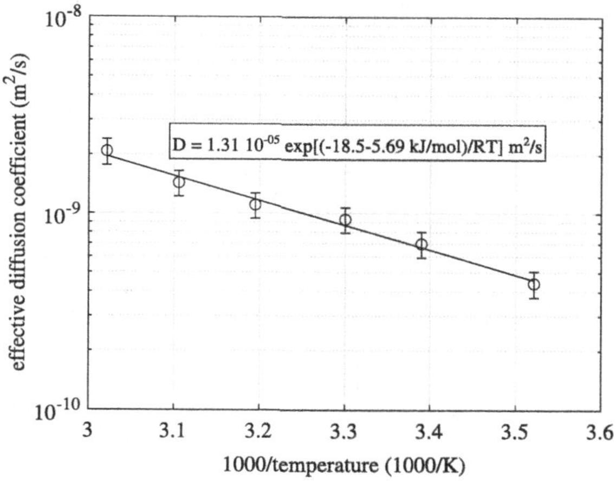

Fig. 12.5 Hydrogen flux varying sample's thickness

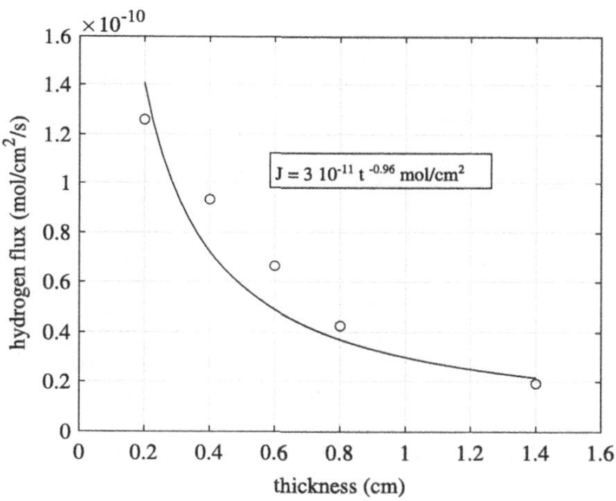

After reaching the maximum (point 1 in Fig. 12.6), the hydrogen signal decreases as an effect of crack nucleation: hydrogen accumulates in the cracks, while the amount exiting from the sheet decreases (point 2); when the new traps are filled the permeation flux starts to increase again and the signal goes up (point 3). Once determined the hydrogen concentration in steel, correspondent to the maximum peak of Fig. 12.6 (5.1 ppmw), and with reference to Fig. 12.7, the hydrogen accumulation next to the crack can be calculated. Considering the theoretical profile of hydrogen concentration along with the specimen thickness during permeation, [21] (Fig. 12.7(a)), the relationship between thickness and hydrogen concentration can be assumed linear:

$$C_H = C_{H,0}\left(1 - \frac{x}{L}\right) \quad (12.12)$$

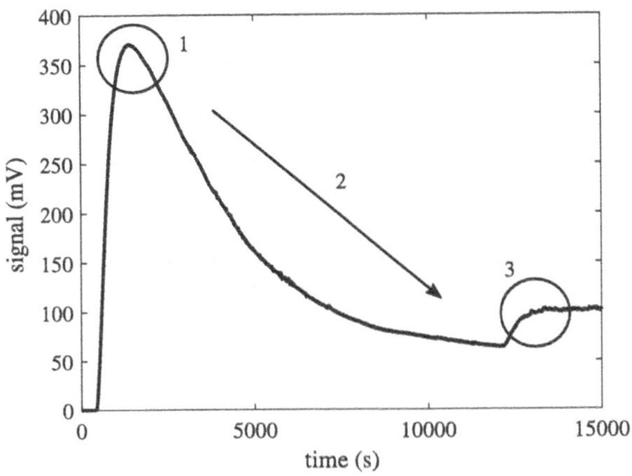

Fig. 12.6 Permeation in crack condition

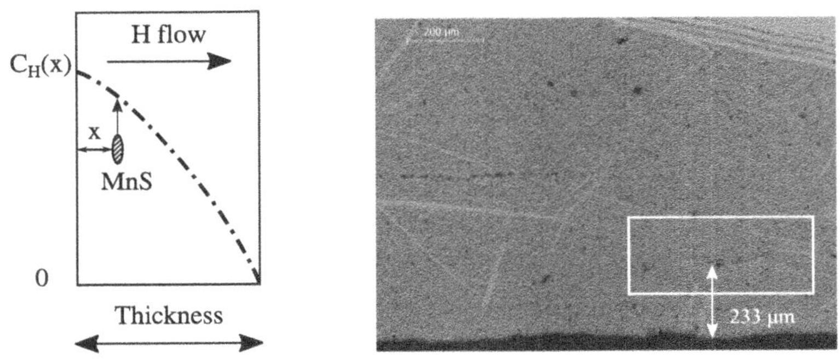

(a) (b)

Fig. 12.7 **a** Hydrogen concentration profile and **b** crack and its distance from sample surface

where $C_{H,0}$ is the hydrogen concentration on the input surface, 5.1 ppmw (found from the HIC tests, where the sample was saturated, so the surface concentration is equal to the core one); x is the distance shown in Fig. 12.7(b) and L the sample thickness (2 mm). Given these considerations, C_H was estimated equal to 4.6 ppmw, and the result was in accordance with HIC results, shown in the following paragraphs.

Figure 12.8 shows the permeation curve of 2 mm thickness sample returned by HELIOS 2. The outcoming hydrogen flux, measured by the device, is represented as a function of time. Signal steady state is equal to $9.4 \cdot 10^{-6}$ mol/s m^2 and the curve was also used to identify the diffusion parameters through the previously exposed numerical model.

Fig. 12.8 Permeation curve obtained using HELIOS 2 on a 2mm thick sample

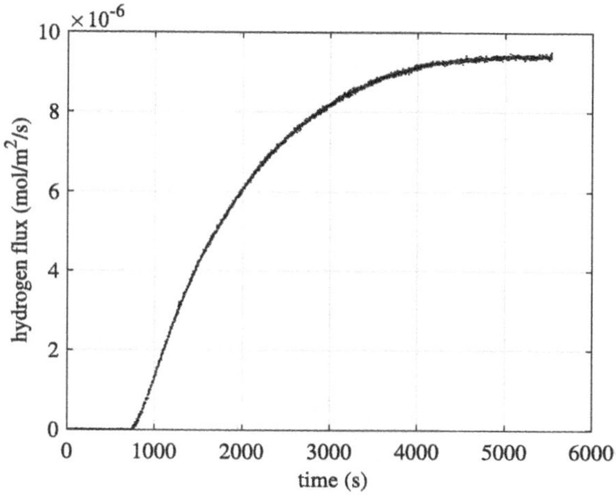

Fig. 12.9 TPD test results

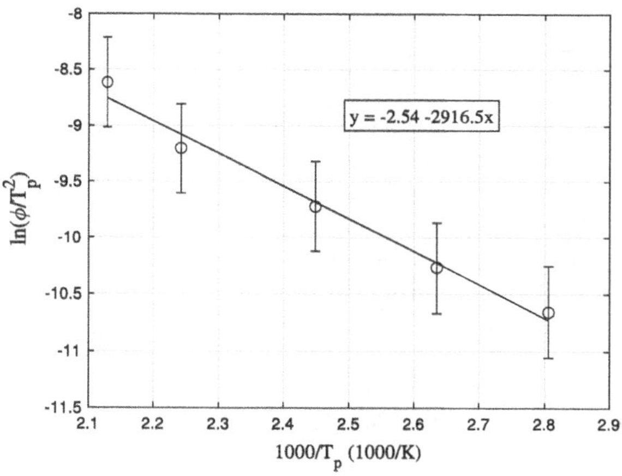

12.4.2 TPD Tests

Different heating rates have been set for TPD analysis: 2, 5, 10 and 20 °C/min. The results were plotted in Fig. 12.9, according to the model of Choo et Lee [13].

According to paragraph 12.3.1, starting from the slope of regression line, the binding energy can be easily estimated and it was equal to 24 kJ/mol.

Fig. 12.10 Results of HIC tests on API X60 steel

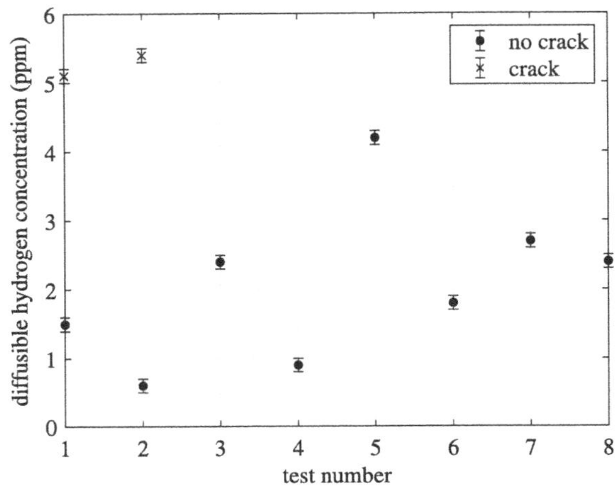

12.4.3 *HIC Tests*

HIC tests were carried out on cylindrical samples, after electrochemical hydrogen charging. Coupons were both observed by metallographic analysis to check the presence of cracks and in parallel the hydrogen content was determined by TDA.

The critical hydrogen concentration was estimated to be equal to 4.5 ppmw: this is the hydrogen amount able to crack the material without external load or applied stresses (Fig. 12.10).

12.4.4 *Numerical Analysis*

Starting from experimental data on API X60 steel with thickness 2 mm, a series of studies were carried out by numerical analysis. Figure 12.11 shows the comparison between the permeation experimental curve and the simulated one. The model allowed to calculate the occupancy of reversible traps. Figure 12.12 shows instead the simulated percentage occupancy of reversible traps as a function of the spatial coordinate for different permeation times. As can be seen, the occupancy of reversible traps remains small over time. The fractional occupation of the trap much smaller than one during the permeation test. Table 12.4 shows the diffusion and trapping parameters obtained by the simulation.

Equation (12.13) is a kinematic expression of Oriani's effective diffusion coefficient derived from the local equilibrium between atomic populations in the normal lattice and traps when the occupancy of trapping sites is very low [22].

$$D_{\text{eff}} = D\left(1 + N_r \frac{k_r}{p}\right)^{-1} \tag{12.13}$$

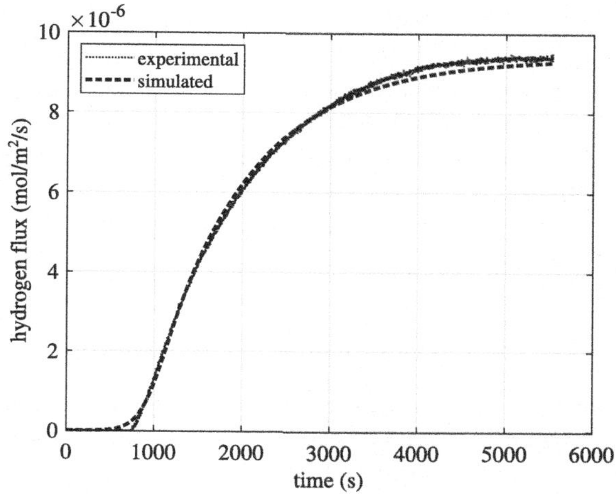

Fig. 12.11 Permeation curve. Comparison with the simulated one

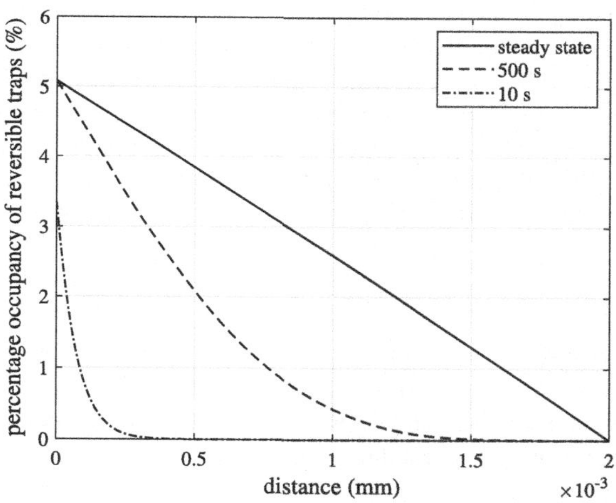

Fig. 12.12 Simulated percentage occupancy of reversible traps at different test times

Table 12.4 Diffusion and trapping parameters identified by the permeation test

C_0 (mol/m^3)	N_r (mol/m^3)	N_i (mol/m^3)	k_r (m^3/(mol s))	k_i (m^3/(mol s))	p (1/s)
2.6	$9.1 \cdot 10^2$	1.3	$2.4 \cdot 10^{-3}$	$2.0 \cdot 10^{-1}$	$1.1 \cdot 10^{-1}$

The reversible trap concentration calculated using the Oriani model leads to a value equal to 210 mol/m^3. This value is of the same order of magnitude as that identified by the numerical model.

12.5 Discussion

Pd coating is typically applied for Devanathan–Stachurcki permeation tests, it avoids the recombination of hydrogen atoms into molecular hydrogen before their oxidation in H^+ (electron release), with consequent reduction of detected current in the reading semi-cell.

In this case, the hydrogen flux exiting from the sample surface can be calculated as follows:

$$J_{De} = \frac{i}{F} \tag{12.14}$$

where i is the current density and F the Faraday's constant. Considering the results of Table 12.3, it can be noticed that without the Pd coating the underestimation of hydrogen flux and concentration is not negligible.

HELIOS 2 working principle is different and there is no need for Pd surface deposition since hydrogen escaping out of the metallic sample after recombination into molecular H_2 reacts with a solid-state gas sensor. This case shows the variation of sensor resistivity which returns an electrical signal in volts. Translating the signal into a hydrogen concentration and flux, the comparison with Devanathan method was possible and is reported in Fig. 12.3i

The flux through HELIOS 2 probe was calculated after a calibration of the instrument, evaluating molar amount per second in stationary conditions:

$$J_{He} = \frac{n}{A} \tag{12.15}$$

with A is the sample area in contact with the probe. HIC tests highlighted a good resistance to hydrogen damage of API X60 steel. Even though the quality of steel manufacturing can play a crucial role, a critical value near 4.5 ppmw was found. The result was also validated by the permeation test in severe environment, where the hydrogen flux was registered, and crack nucleation was identified through the sensor signal progress.

12.6 Conclusion

Devanathan-Stachurski method can be used for the calculation of hydrogen diffusion coefficient, but the Palladium coating deposition is needed for a reliable estimation of hydrogen flux coming out of the sample, especially for high values of hydrogen flux. This pointed out the elevated precision of the double cell for laboratory investigation but also the impossibility to apply the method to a real pipeline.

HELIOS technique, based on a solid-state gas sensor, is demonstrated to be very practical. Thanks to the high sensitivity device it can be applied directly to the pipe wall. The laboratory investigation carried out in the present study, was aimed at approaching and validating the first step toward online monitoring.

The measurement carried out with HELIOS II of the hydrogen outlet flux was interpreted through a hydrogen diffusion and trapping numerical model providing results in line with the classical theory.

However, the realization of permeation tests with Devanathan-Stachurski on Pd coated specimens, can be successfully used for calibration and validation alternative methods.

Through the TPD and HIC tests, it has been possible a complete characterization of the material in terms of hydrogen trapping and binding energy [23] and the critical hydrogen concentration giving rise to cracks in the material.

Furthermore, the validation of new sensors and hydrogen evaluation suitable for the industrial application have been carried out and achieved.

References

1. Scopus. Elsevier, Amsterdam. http://www.scopus.com (2004). Accessed 16 June 2020
2. Yang, Y., Khan, F., Thodi, P., Abbassi, R.: Corrosion induced failure analysis of subsea pipelines. Reliab. Eng. Syst. Safety. **159**, 214–222 (2016)
3. Djukic, M., Bakic, G., Zeravcic, V., Sedmak, A., Rajicic, B.: The synergistic action and interplay of hydrogen embrittlement mechanisms in steels and iron: localized plasticity and decohesion. Eng. Fract. Mech. **216**, 106528 (2019)
4. Katzarov, I., Pashov, D. L., and Paxton, A.: Hydrogen embrittlement I. Analysis of hydrogen-enhanced localized plasticity: effect of hydrogen on the velocity of screw dislocations in α-Fe. Phys. Rev. Mater. **1**, 1–11 (2017)
5. Taketomi, S., Matsumoto, R., Hagihara, S.: Molecular statics simulation of the effect of hydrogen concentration on 112 <111>edge dislocation mobility in Alpha Iron. ISIJ Int. **57**, 2058–2064 (2017)
6. Gerberich, W., Stauffer, D. and Sofronis, P.: A coexistent view of hydrogen effects on mechanical behaviour of crystals: HELP and HEDE effects of hydrogen on materials. In: Somerday, B., Sofronis, P.,Jones, P.(eds.) Effects of hydrogen on materials: proceedings of 2008 international hydrogen conference Jackons Lake Lodge, Grand Teton National Park, Wyoming, USA (2008)
7. Peng, Z., Liu, J., Huang, F., Hu, Q., Cao, C., Hou, S.: Comparative study of non-metallic inclusions on the critical size for HIC initiation and its influence on hydrogen trapping. Int. J. Hydrogen Energy **45**, 12616–12628 (2020)
8. Pressouyre, G., Bernstein, I.: An example of the effect of hydrogen trapping on hydrogen embrittlement. Metall. Trans. A **12A**, 835–844 (1981)
9. Fragiel, A., Serna, S., Malo-Tamayo, J., SIlva, P., Campillo, B., Martinez-Martinez, E., Cota, L., Staia, M., Puchi-Cabrera, E. and Perez, R.: Effect of microstructure and temperature on the stress corrosion cracking of two microalloyed pipeline steels in H2S environment for gas transport. Eng. Fail. Anal. **105**, 1055–1068 (2019)
10. ISO 17081:2014: Method of measurement of hydrogen permeation and determination of hydrogen uptake and transport in metals by an electrochemical technique. ISO, Geneva (2004)
11. Manolatos, P., Jerome, M.: A thin palladium coating on iron for hydrogen permeation studies. Electrochim. Acta **41**, 359–365 (1996)
12. Valentini, R.: Methods for permeation hydrogen measurement. Patent EP2912452, 7 Dec 2012
13. Choo, W., Lee, J.: Thermal analysis of trapped hydrogen in pure iron. Metall. Trans. A **13A**, 135–140 (1982)

14. Pourazizi, R., Mohtadi-Bonab, M., Szpunar, J.: Role of texture and inclusions on the failure of an API X70 pipeline steel at different service environments. Mater. Charact. **164**, 110330 (2020)
15. Manolatos, P., Jerome, M., Duret-Thual, C., Le Coze, J.: The electrochemical permeation of hydrogen in steels without palladium coating. Part I: interpretation difficulties. Corros. Sci. **37**, 1773–1783 (1995)
16. Mehrer, H.: Diffusion in solids. Springer, Berlin (2007)
17. McNabb, A., Foster, P.K.: A New Analysis of the diffusion of hydrogen in iron and ferritic steels. Trans. Metall. Soc. **227**, 618–627 (1963)
18. Hagi, H., Hayashi, Y., Ohtani, N.: Diffusion coefficient of hydrogen in pure iron between 230 and 300K. Trans. Jpn. Inst. Met. **20**, 349–357 (1979)
19. Nagumo, M.: Fundamentals of hydrogen embrittlement. Springer, Singapore (2016)
20. Skjellerudsveen, M., Akselsen, O., Olden, V., Johnsen, R., Smirnova, A.: Effect of microstructure and temperature on hydrogen diffusion and trapping in X70 grade pipeline steel and its weldments. In: Presented at the second European Corrosion Congress, vol. 13–17. Moscow (September 2010)
21. Crank, J.: The mathematics of diffusion. Oxford University Press, London and New York (1957)
22. Oriani, R.: The diffusion and trapping of hydrogen in steel. Acta Metal. **18**, 147–157 (1970)
23. Villa, M.: Controllo dei fenomeni di HE nei tubi in acciaio per oil and gas: applicazione di sistemi innovativi di monitoraggio. Dissertation, University of Pisa (2018)

Chapter 13
Hydrogen Diagnostics of Industrial Parts of Aluminum Alloys

Yuriy A. Yakovlev, Dmitriy G. Arseniev, Alexander K. Belyaev, Victor P. Loginov, Anatoliy M. Polyanskiy, and Vladimir A. Polyanskiy

Abstract Aluminum alloys are very popular in a variety of technical applications. The strong influence of hydrogen on the properties of aluminum alloys is known, however, as in the case of steels; it is continuously increasing as new alloys with extreme properties are developed and introduced. Scientific research in the field of the hydrogen effect on the properties of aluminum alloys is mainly focused on the fundamental aspects such as the diffusion coefficients of hydrogen in aluminum, possible types of hydrogen traps, and their effect on the microstructure of alloys. At the same time, the industry has a problem of cracking ingots and semi-finished products (sheets, pipes, and plates), including their further processing and welding. In contrast to the high-strength steels, scientific research does not actually provide specific values for critical hydrogen concentration. The problem of separating the hydrogen adsorbed on the surface and dissolved during measurements has not been solved. There are only a few types of aluminum alloy reference specimens. The article is intended to partially fill this gap. It provides specific examples of the study of technological problems and proposes the measurement methods that allow the separation of hydrogen dissolved and adsorbed on the surface.

Keywords Hydrogen analyzer · Aluminum alloys · Surface hydrogen · Measurement of hydrogen concentration

Y. A. Yakovlev (✉) · A. K. Belyaev · V. P. Loginov · V. A. Polyanskiy
Institute for Problems in Mechanical Engineering RAS, Bolshoy pr., 61, 199178 St. Petersburg, VO, Russia
e-mail: yura.yakovlev@gmail.com

D. G. Arseniev
Peter the Great St. Petersburg Polytechnic University (SPbPU), Polytechnicheskaya str. 29, 195251 St. Petersburg, Russia

A. M. Polyanskiy
LLC «RDC Electronic and Beam Technologies», Politekhnicheskaya str. 28, lit A, 194064 St. Petersburg, Russia

V. A. Polyanskiy and A. K. Belyaev (eds.), *Advances in Hydrogen Embrittlement Study*, Advanced Structured Materials 143, https://doi.org/10.1007/978-3-030-66948-5_13

13.1 Introduction

Aluminum alloys are widely used in all areas of technology, from street lighting to space systems.

Hydrogen appears in aluminum alloys during their manufacture due to the reaction of aluminum with water and therefore is a critical element of these alloys. The presence of dissolved hydrogen completely determines the porosity of aluminum alloys [1–8]. Hydrogen leads to hydrogen embrittlement, crystallization cracking of ingots, environment-assisted cracking [9, 10], and a decrease in fatigue strength [10, 11].

Much attention in the published articles is paid precisely to the study of various aspects of the interaction of hydrogen with aluminum and its influence on the microstructure of aluminum-based alloys.

In Refs. [12–14], on the basis of experimental data, various mechanisms of the formation of vacancies and pores in aluminum during its crystallization from the melt are discussed. The paper [15] discusses the mechanisms of hydrogen diffusion on the surface and inside an aluminum single crystal. An increase in the concentration of vacancies and pores and the effect of dissolved hydrogen are modeled on the basis of the first principles in [1]. Reference [16] is concerned with the nature and density of hydrogen traps in Al–Zn–Mg alloys, while [17] experimentally investigated the mechanism of pore formation in AlSi7Mg0.3 ingots.

Microdefects arising in a single crystal of aluminum and polycrystalline aluminum are investigated when the surface of samples is bombarded with proton beams and interacts with hydrogen plasma; see [18, 19]. An increase in the concentration of vacancies and the main energy relationships of this process are discussed too.

Micromechanisms of hydrogen traps in aluminum alloy 6061-T6 are investigated in [20].

The effects of cathodic hydrogen charging of aluminum alloy samples were studied in [21, 22].

The temperature dependence of the hydrogen diffusion coefficient was obtained in [23–26].

The limiting solubility of hydrogen in aluminum alloys and the activation energies of hydrogen traps were studied in [27–35].

At the same time, experimental data on the parameters of the interaction of hydrogen with aluminum and its possible states in aluminum alloys published over the past 30 years have a wide spread. For example, the published values of diffusion constant have range $4 \cdot 10^{-6}$–$1 \cdot 10^{-9}$ m^2s^{-1} and diffusion activation energies vary between 0.3 and 1.3 eV, cf. [25, 36–40].

The reasons for this spread may be strong surface effects typical for aluminum alloys [15]. In [41], the influence of the oxide layer and vacuum cleaning of the surface on the permeability of aluminum membranes was studied, and it was concluded that the state of the surface has a large effect on the permeability and the actually measured diffusion coefficient of hydrogen.

Modern aluminum alloys, like high-strength stills, are characterized by the values of the mass fraction of hydrogen at the level of 0.04–0.05 ppm. For aluminum, a hydrogen concentration of 0.1 ppm means that for every molecule of H_2 absorbed by the metal, there are about a million aluminum atoms.

Most of the various studies have been carried out without quantitative measurements of the hydrogen concentrations in aluminum and aluminum alloys and without the discussion of specific values of these concentrations. This fact significantly reduces their practical value. Only in a small number of works, for example in [3, 5, 8, 21, 22, 42–44], the hydrogen concentrations were measured and the dependence of the studied effects on these concentrations was studied. Moreover, in some of these works, concentration ranges were investigated up to 7 ppm [22] or even 400 ppm [21] which is not feasible in practice.

Such a spread in the experimental data on concentrations is explained by the fact that when measuring the hydrogen concentration in aluminum alloys, the hydrogen adsorbed on the surface of samples or "surface" hydrogen [44] is a big hindrance, which introduces significant distortions in the measurement results. The standard [45] establishes rules for the exclusion of surface hydrogen, but only for cylindrical samples of two standardized diameters of 8 and 10 mm. In other cases, there are no approaches to the separation of surface hydrogen from the total flux of hydrogen desorbed from the sample. This makes quantitative measurements much more difficult.

Unlike steels, the effect on the properties of low natural concentrations of hydrogen has been little studied, and it is these concentrations that are most often encountered in practical problems. The proposed article is intended to partially fill this gap.

13.2 Experimental Equipment

For the control and determination of hydrogen concentration, we use the hydrogen analyzer AV-1. It can be used to reliably measure both natural, ultra-low hydrogen concentrations, and after special saturation, when concentrations can reach several thousand ppm. The precision hydrogen analyzer AV-1 has been designed for hydrogen detection in metals and alloys under the plant laboratory conditions during the outgoing moldings control with various alloys [44, 46–48]. The operating principle of the analyzer is mass-spectrometric.

The sample processing is shown in Fig. 13.1, which includes a vacuum extractor and a heater (4). The vacuum extractor is made of quartz glass and has three setoffs. The first setoff (1) contains the samples before the test. The hydrogen content analysis takes place in the second setoff (2). The samples which have been tested are kept in the third setoff (3).

Before testing, the heater (4) with a previously set temperature is placed upon the second setoff (2) of the extractor. With the help of a magnetic push rod, the test sample from the first setoff is thrown into the second analytic setoff of the extractor. During the test, the metal sample within the second setoff is subjected to gradual

Fig. 13.1 The sample processing of hydrogen analyzer AV-1

heating up to the extraction temperature. The extraction temperature for the sample is lower than its flowing temperature, as a rule, it is within the interval of 200–900 °C.

The fumes that have been emitted during heating are evacuated from the extractor before the pressure reaches 50 micro Pascal and are analyzed with a mass spectrometer which is located in the vacuum pumping pass customized for the hydrogen line. The end of the testing is identified with the help of a background signal. The background signal remains constant before the beginning and at the end of the testing. As soon as the level of the signal from the sample becomes equal to the background signal, the test stops and the sample from the analytical setoff (2) is placed into the setoff for the samples which have been analyzed (3). For this purpose, the heater (4) is lowered down with the help of a hoisting gear, and the extractor keeps turning about itself until the sample being tested falls down from setoff (2) into setoff (3) under gravity. After that, the extractor is returned to the original position, and the heater is placed on setoff (2) with the help of a hoisting gear. After setting the background hydrogen value, the test continues for the next sample from setoff (1). In such a manner, the next sample is analyzed without vacuum failure. It should be emphasized that the extraction temperature doesn't change during the test. This makes it possible to reach the high stability of the background signal.

The time dependence of hydrogen flow $q_H(t)$ is recorded by the system of digital registration as an extraction curve. Figure 13.2 shows an extraction curve for aluminum alloy AMg6, which is a simple example to describe.

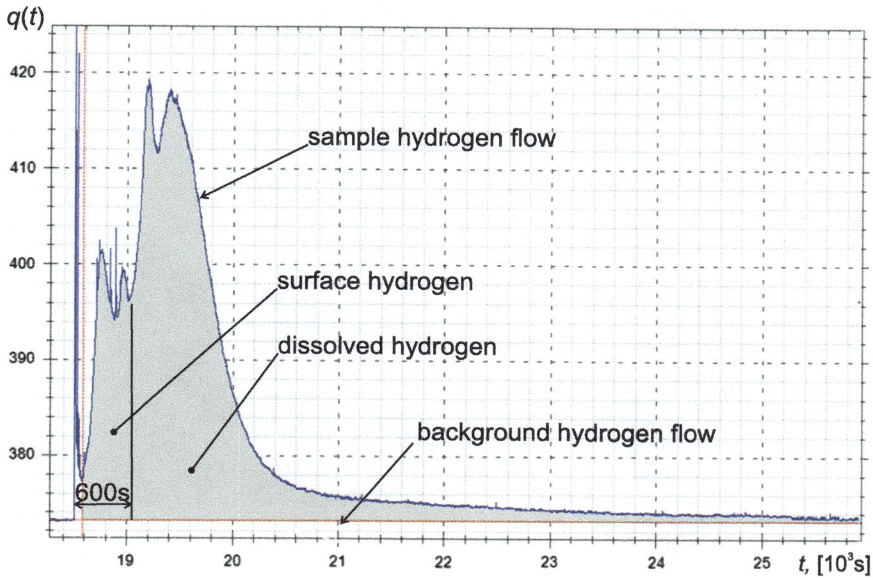

Fig. 13.2 Extraction curve for aluminum alloy AMg6

The time integral from the extraction curve (painted gray in Fig. 13.2) in reference to the background hydrogen flow is proportional to the hydrogen evolved from the sample. The proportionality factor is set during analyzer calibration.

The background hydrogen flux is determined prior to heating the sample. When cylindrical samples with a diameter of 8 mm are used, the first 600 s after the start of heating, hydrogen flux adsorbed on the sample surface or "surface" hydrogen is observed [44, 45]. All subsequent hydrogen flux refers to the extraction of dissolved or internal hydrogen [45].

Each peak of the extraction curve can be correlated with its hydrogen binding energy [44].

13.3 Investigation of the Causes of Rupture of Aluminum Tubes

Aluminum tubes are made by impact extrusion. A striker made of high-strength steel at high speed hits a round plate which is a rondole (tablet) of high-purity aluminum alloy. It causes high-speed plastic deformation of the tablet in the mold and the formation of all parts of the tube from one workpiece. High plastic deformations of thousands of percent are observed. The tablets and produced tubes are shown in Fig. 13.3.

Fig. 13.3 Aluminum tablets and produced tubes

Table 13.1 Composition of alloy A7 (%)

Fe	Si	Ti	Al	Cu	Zn
<0.16	<0.16	<0.02	99.7	<0.01	<0.04

Fig. 13.4 Scheme of cutting samples from tablets

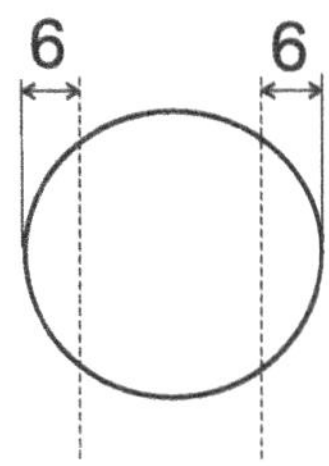

With this method of tube production, a significant number of ruptures were observed on the tube surface, shown in Fig. 13.3. The wall thickness of the tube is less than 1 mm. In this case, strong surface effects are observed associated with the hydrogen sorption on the surface and the non-uniform formation of the oxide film; see [41]. In this regard, it was decided to conduct a study of a batch of the material of tablets with a diameter of 23 mm and a thickness of 3 mm made of alloy A7 with aluminum 99.7%. The alloy composition is shown in Table 13.1.

Two specimens were cut from the tablets as shown in Fig. 13.4.

Standard measurements of hydrogen concentration by hot vacuum extraction were carried out. A representative extraction curve is shown in Fig. 13.5.

Surface hydrogen was separated by the extraction curve according to [44, 45]. In this case, only two peaks of the extraction curve are observed, moreover, the samples

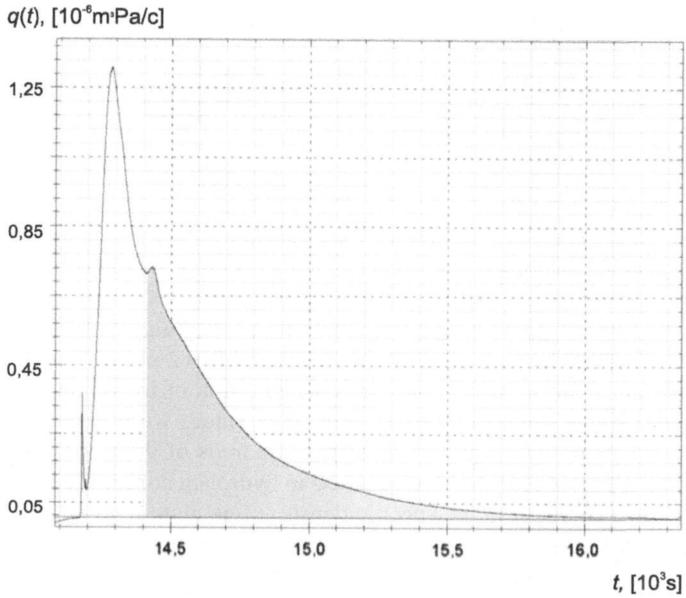

Fig. 13.5 Extraction curve for tablets

Table 13.2 Results of measuring the concentration of hydrogen in tablets. Q_s and Q_d denote surface and dissolved hydrogen, respectively

#	Q_s, [ppm]	Q_d, [ppm]
1	1.407	0.883
2	0.376	0.581
3	1.219	0.934
4	0.435	0.613
5	0.923	0.960
6	0.575	0.821
7	0.616	0.887
8	0.552	0.814
9	0.436	0.640
10	0.546	0.683

have a characteristic size close to the standard and the duration of the first peak of ca. 500s is slightly less than the standard time of extraction of the surface hydrogen. In this case, the standard method, when only the first peak is associated with surface hydrogen, gives an adequate result.

The data averaged over two samples for each tablet are presented in Table 13.2.

Table 13.3 Composition of alloy AMg6 (%)

Fe	Si	Mn	Ti	Al	Cu	Be	Mg	Zn
<0.4	<0.4	0.5–0.8	0.02–0.1	91.1–93.68	<0.1	0.0002–0.005	5.6–6.8	<0.2

One can single out the main factor that speaks of the low quality of the table metal: all the tablets belong to the same batch, that is, they are made from one ingot, but there is a strong spread in the measured hydrogen concentrations. The minimum values of $Q_d = 0.581$ ppm are approximately four times higher than the hydrogen concentration typical for this alloy (in the range of 0.1–0.2 ppm).

An important point is that the normal concentration of hydrogen in aluminum, as a rule, lies in the range of 0.06–0.2 ppm. The values we measured are several times higher, but still make up one atom per hundreds of thousands of aluminum atoms. As the data from [5] show, an increase in hydrogen concentration by 0.1 ppm corresponds to a decrease in the specific density of an aluminum alloy by 10–20% due to the growth of pores. Thus, it can be concluded that, in this case, the excess hydrogen porosity led to a loss of tube wall continuity during extrusion.

The given example shows that the technique of surface hydrogen evolution is the most important element of the metrological system. Without this technique, it is not possible to obtain a reliable quantitative result when measuring the hydrogen concentration in aluminum alloys, since the hydrogen adsorbed on the surface can significantly exceed the internal one.

13.4 Investigation of the Causes of Pipe Cracking During Plastic Deformation

We analyzed samples of aluminum pipes of the AMg6 alloy with a diameter of 120 and 160 mm and a wall thickness of 7 mm, from which the high-pressure gas cylinders are made by pressing. Samples of two types of pipes were taken. Samples of the first type belonged to a batch of pipes from which high-quality cylinders were obtained. Samples of the second type were cut from a batch of pipes, from which the cylinders are obtained with cracks in the walls. The alloy composition is shown in Table 13.3.

To determine the hydrogen content, short annular sections with a length of 15 mm were cut off from the pipes. They were cut along the generatrix of the pipe, according to the diagram in Fig. 13.6.

In all pipes, which are prone to cracking when the cylinders are pressed from them, there is an increased average concentration of dissolved hydrogen of 0.65 ppm and a large spread of the results of individual measurements in each ring (Fig. 13.6) from 0.4 ppm to 0.95 ppm. Please note that these were samples cut with an interval of 10–20 mm between samples.

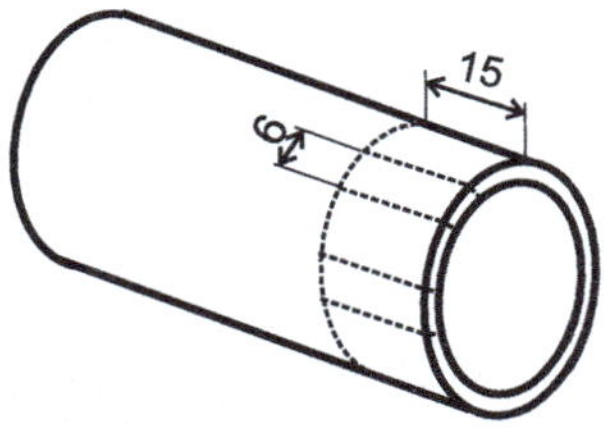

Fig. 13.6 Scheme of cutting samples from a pipe

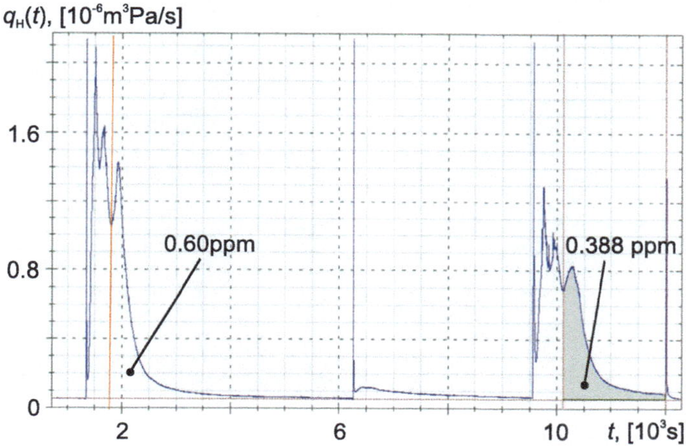

Fig. 13.7 Extraction curves of samples of alloy AMg6

The average hydrogen content of samples of the first batch of proper quality was 0.38–0.39 ppm. The evolution of surface hydrogen remains an important problem. Figure 13.7 shows the extraction curves obtained by analyzing the hydrogen content for these samples. Three peaks are clearly visible on the curves, according to the requirements [45]; only the first peak is related to surface hydrogen. But if we focus on the characteristic time of desorption of the surface hydrogen from samples with a characteristic size of 6–8 mm (about 10 min), then the first two peaks must be attributed to the surface hydrogen.

Figure 13.7 on the left shows the extraction curve for a sample from a low-quality tube, and on the right the curve for a normal sample is shown. With the general similarity of the curves (the number of maxima and their location), it is clearly seen that the area under the first curve is noticeably larger than the area under the second curve.

The area under the extraction curve is proportional to the hydrogen concentration. Taking into account the mass of the sample, the concentration of dissolved hydrogen differs by 2.5 times (the third peak in each of the curves in Fig. 13.7).

Table 13.4 Alloy composition of AD0

Fe	Si	Mn	Ti	Al	Cu	Mg	Zn
<0.3	<0.3	<0.025	<0.1	99.5	<0.15	<0.03	<0.07

Fig. 13.8 Samples for analysis from alloy AD0

Such a spread in hydrogen concentration can be explained by the increased porosity or the presence of the non-uniform distribution of magnesium over the volume of the alloy. Recently, an increased hydrogen concentration associated with a significant proportion of non-metallic inclusions has been increasingly encountered.

13.5 Investigation of the Causes of Poor Weldability of AD0 Alloy Sheets

Foaming and subsequent cold cracking of the weld were observed when welding AD0 alloy sheets. Samples for analysis shown in Fig. 13.8 8 × 15 mm in size were cut from a sheet of 10 mm thickness. The photograph shows that the entire surface of the samples comes out with relatively large pores. The alloy composition is shown in Table 13.4.

When analyzing these samples, one observes numerous peaks on the extraction curve; see Fig. 13.9. When pores are emptied, a sharp release of hydrogen occurs into the vacuum system of the AV-1 hydrogen analyzer, which is recorded by the mass spectrometer as narrow peaks of large amplitude.

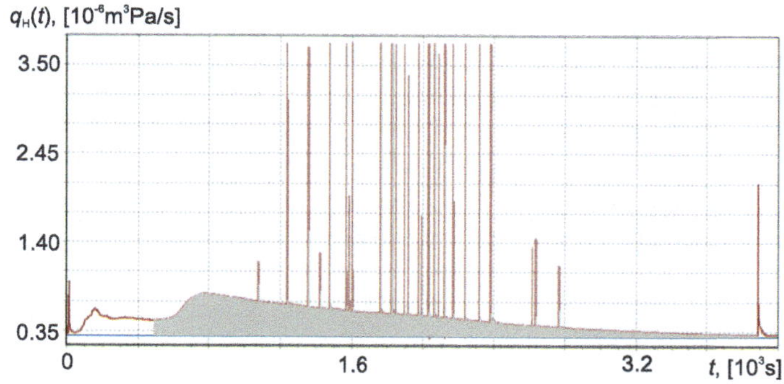

Fig. 13.9 Extraction curve of the sample of aluminum alloy AD0

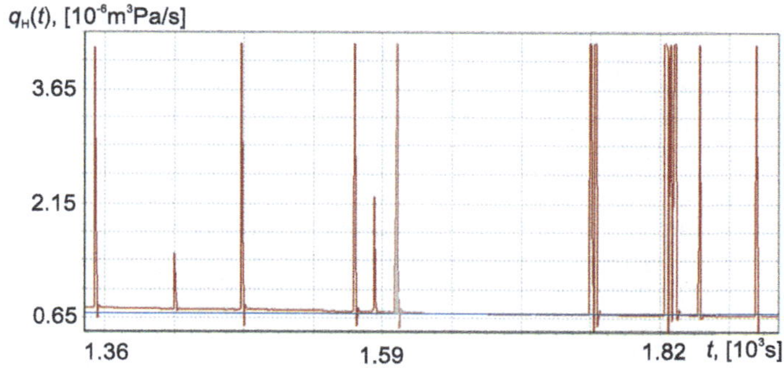

Fig. 13.10 An enlarged fragment of the extraction curve alloy AD0

An enlarged fragment of the extraction curve in Fig. 13.10 demonstrates that the release of hydrogen from the pores occurs smoothly over a finite time interval of 5–10 s. This makes it possible to reliably determine the hydrogen content at such maxima.

The entire curve is calibrated against standard samples for measuring on AV-1; thus, it is possible to calculate the hydrogen amount corresponding to the concentration of hydrogen in each peak. Under normal conditions, it corresponds to a volume of hydrogen of 0.125 norms mm^3. If we assume that the hydrogen in the pore is at approximately atmospheric pressure and that the pore is spherical, then its diameter will be 0.6 mm. The pores of exactly this size, coming out to the surface, are clearly visible in Fig. 13.6. Thus, 20–30 such pores contain the amount of hydrogen, which increases the total hydrogen concentration in the metal by 0.1–0.2 ppm. This effect was observed in the samples studied, where the average hydrogen concentration was 0.4 ppm, that is, it was only twice the normal concentration of dissolved hydrogen for AD0.

Consequently, even a twofold excess of the hydrogen concentration significantly increases the porosity of aluminum alloys, which can lead to cracking and hydrogen embrittlement during the subsequent processing and operation of the metal.

The hydrogen content is controlled for the most critical industries, but only in an ingot. In various technical processes, the concentration of hydrogen can both increase and decrease. Therefore, it is necessary to control the hydrogen content in semi-finished products. At the moment, there is no methodological basis for such control.

13.6 Examination of Samples from AlMn Sheet

Plastic deformation leads to significant measurements in the distribution of dissolved hydrogen concentrations [49, 50]. At the same time, the methodological issue of the separation of surface hydrogen becomes especially important, since during rolling there is a significant deformation of the metal surface in a heated state, which should change not only the distribution of dissolved hydrogen, but also the distribution of surface hydrogen.

For the experiments, 10 samples were taken, which were cut from a rolled sheet with a thickness of 15 mm of the AlMn alloy. The alloy composition is shown in Table 13.5.

The cutting diagram is shown in Fig. 13.11.

Two prismatic specimens were cut from the resulting prismatic specimens, as shown in Fig. 13.12. The dimensions are based on material loss per cut width. A layer

Table 13.5 Composition of the AlMn alloy

Fe	Si	Mn	Ti	Al	Cu	Mg	Zn
<0.7	<0.6	1–1.6	<0.2	96.35–99	< 0.15	< 0.2	<0.1

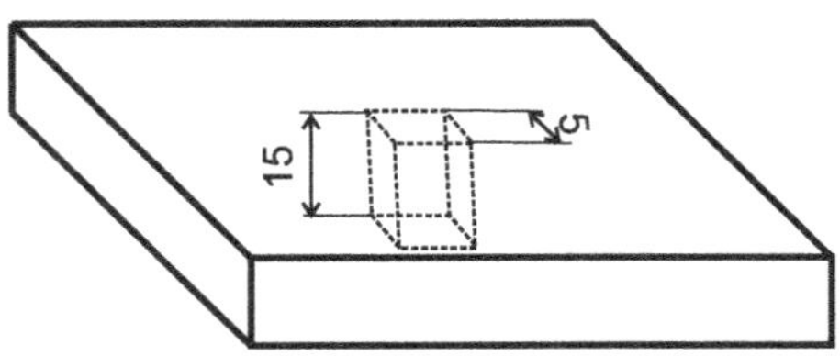

Fig. 13.11 Scheme of cutting samples from a sheet

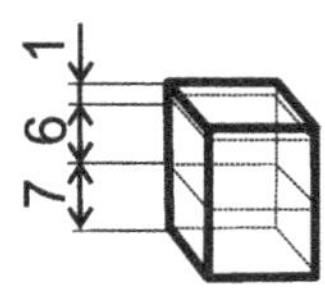

Fig. 13.12 Sample cutting scheme for measurements

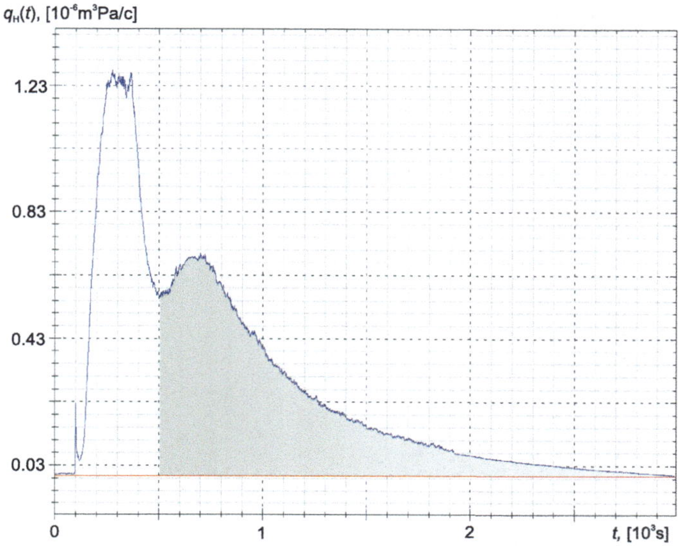

Fig. 13.13 Extraction curve of a sample with a rolling layer. The area corresponding to dissolved hydrogen is highlighted in gray

of the rolling surface of 1 mm thickness was removed from one of the samples by grinding; see Fig. 13.12.

Measurements of hydrogen concentration were carried out on AV-1 by the vacuum heating method in a standard way. Two typical extraction curves are shown in Figs. 13.13 and 13.14.

Surface hydrogen was released in accordance with the procedure [44, 45]. The concentration of dissolved hydrogen in the samples with the removed rolling layer was from 0.185 to 0.215 ppm. The concentration of dissolved hydrogen in the samples with the rolled bed was from 0.46 to 0.64 ppm. As is clearly seen in Figs. 13.13 and 13.14, the amount of surface hydrogen differs by about tens of times, but the desorption time is the same (400 s) which is due to the smaller size of the samples as compared to the standard 8 mm size.

One of the results of this study is a multiple increase in the concentration of dissolved hydrogen in the surface layer of the metal as a result of deformation during rolling. By comparison, the concentration value can be estimated at 2.45 ppm, which is 12 times higher than the values in the middle part of the sample. This is one of their manifestations of the surface effect during plastic deformation, which we have already described in [49, 50]. The data obtained also indicate that the desorption time of surface hydrogen depends on the sample size, which must be taken into account when measuring the hydrogen concentration in thin samples with a characteristic thickness of less than 7–8 mm.

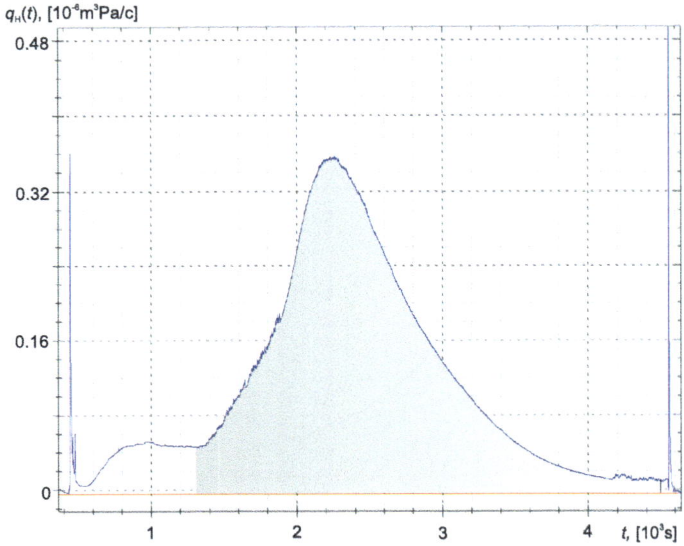

Fig. 13.14 Extraction curve of a sample with a removed rolling layer. The area corresponding to dissolved hydrogen is highlighted in gray

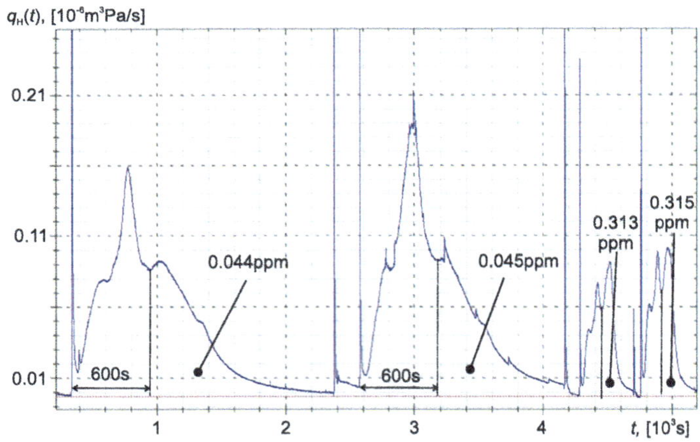

Fig. 13.15 Extraction curves of the AMg6 alloy samples: 2 curves on the left are for a bar and 2 curves on the right are for a wire

13.7 Release of Surface Hydrogen in Thin Samples

Measurements of the hydrogen concentration were carried out in samples of AMg6 alloy, machined from a 9 mm diameter rod and a 1.2 mm diameter wire. The wire was obtained by extrusion from rods of the same batch as those studied. This wire is used in automatic welding. We used samples D7 × 15 mm^2 turned from a bar on a lathe and samples 15 mm long without surface treatment, cut from the wires. Figure 13.15 shows the extraction curves for two bar samples and two wire samples. In a time interval of 10 min, the first two peaks of the extraction curve appear in the samples from the bar; see Fig. 13.15.

The extraction curves of thin wire samples are also characterized by three peaks, and the first two peaks should probably be attributed to surface hydrogen (focusing on more bulk samples). Taking these assumptions into account, the hydrogen concentration we measured in the 9 mm diameter AMg6 alloy was from 0.04 to 0.07 ppm, and in the wire extruded from this rod it was from 0.31 to 0.33 ppm.

It should be noted that during extrusion, the hydrogen concentration increased fourfold, remaining in the usual range of natural hydrogen concentrations for an alloy of this type. This result should be taken into account, since it is generally accepted that the hydrogen concentration during production from an ingot to finished product cannot change significantly due to the low diffusion mobility of hydrogen in aluminum. Therefore, the control over the concentration of hydrogen is carried out only at the stage of manufacturing ingots. Note that the first two peaks of the extraction curve responsible for surface hydrogen become noticeably smaller after extrusion, which is probably related to the properties of the wire surface and requires additional structural studies.

13.8 Conclusions

Nowadays, the industry is increasingly faced with problems caused by the effect of hydrogen on the structure and strength of metals and alloys. The effect of dissolved hydrogen in aluminum alloys and high-strength steels is especially strong. The concentrations critical of the properties of these metals are much less than 1 ppm.

A significant contribution to the results of measurements by standard methods of hydrogen adsorbed on the surface is characteristic for aluminum alloys.

The studies described in the present article show that the critical hydrogen concentrations for most alloys are of the order of 0.2–0.4 ppm; their excess by two times leads to rupture and cracking during processing, welding, and excessive porosity.

We have proposed a technique for separating surface hydrogen which has been tested on specific examples and gives adequate results. Further verification and metrological testing of the method are required as well as the determination of the possibility of its application for a wide range of aluminum alloys.

It is generally accepted that the hydrogen concentration does not change during technological processes, that is, the hydrogen concentration in the casting is equal to its concentration in the finished product. Our experiments have shown that this is not the case. During extrusion and in rolled products, the concentration of dissolved hydrogen can increase significantly, but this fact is in no way taken into account by the modern system of standards regulating the maximum permissible hydrogen concentrations, including the thin rolled materials. At the moment, this concentration is assumed to be equal to the concentration of hydrogen in the metal of the original ingot.

One of the reasons for the modern hydrogen problems in aluminum alloys is the markedly increased proportion of secondary metal in finished products. At the same time, standard measurements of hydrogen concentration from two or three hydrogen samples do not guarantee the quality of the material. Our experiments show that the spread in the values of hydrogen concentrations reaches 3–4 times in samples cut from one section of the product with an interval of 15–20 mm.

The quantitative data on hydrogen concentrations available in the modern literature are extremely scanty, do not contain a description of how the surface hydrogen was removed, and this raises many questions, especially if the indicated concentrations significantly exceed 1 ppm. Probably, it is necessary to expand the research, which is usually carried out for steels, to the aluminum alloys and to determine the ranges of safe and dangerous concentrations in terms of structure and strength of the metal more precisely.

Acknowledgements The study has been undertaken with the support of Russian Foundation for Basic Research grants: No-20-08-01100, 18-08-00201.

References

1. Liu, P., Matsumoto, R.: Temperature dependence of vacancy concentration and void growth mechanism in al with constant hydrogen concentration: A first-principles study. Eng. Fract. Mech. **216**, 106508 (2019). https://doi.org/10.1016/j.engfracmech.2019.106508
2. Toda, H., Hidaka, T., Kobayashi, M., Uesugi, K., Takeuchi, A., Horikawa, K.: Growth behavior of hydrogen micropores in aluminum alloys during high-temperature exposure. Acta Mater. **57**(7), 2277 (2009). https://doi.org/10.1016/j.actamat.2009.01.026
3. Thomas, P., Gruzleski, J.: Threshold hydrogen for pore formation during the solidification of aluminum alloys. Metall. Trans. B **9**(1), 139 (1978)
4. Bhuiyan, M.S., Tada, Y., Toda, H., Hang, S., Uesugi, K., Takeuchi, A., Sakaguchi, N., Watanabe, Y.: Influences of hydrogen on deformation and fracture behaviors of high Zn 7xxx aluminum alloys. Int. J. Fract. **200**(1–2), 13–29 (2016)
5. Li, X., Cao, Y., He, L., Guo, Y., Cui, J.: Relationships between casting parameters and inclusions on the hydrogen content in 1050 alloy. Steel Res. Int. **84**(12), 1223–1229 (2013)
6. Atwood, R.C., Sridhar, S., Zhang, W., Lee, P.D.: Diffusion-controlled growth of hydrogen pores in aluminium-silicon castings: in situ observation and modelling. Acta Mater. **48**(2), 405–417 (2000). https://doi.org/10.1016/S1359-6454(99)00363-8
7. Toda, H., Minami, K., Koyama, K., Ichitani, K., Kobayashi, M., Uesugi, K., Suzuki, Y.: Healing behavior of preexisting hydrogen micropores in aluminum alloys during plastic deformation. Acta Mater. **57**(15), 4391–4403 (2009). https://doi.org/10.1016/j.actamat.2009.06.012

8. Lee, P.D., Hunt, J.D.: Hydrogen porosity in directional solidified aluminum copper alloys: in situ observation. Acta Mater. **45**(10), 4155–4169 (1997). https://doi.org/10.1016/S1359-6454(97)00081-5
9. Bond, G.M., Robertson, I.M., Birnbaum, H.K.: The influence of hydrogen on deformation and fracture processes in high-strength aluminum alloys. Acta Metall. **35**(9), 2289–2296 (1987). https://doi.org/10.1016/0001-6160(87)90076-9
10. Song, R.G., Tseng, M.K., Zhang, B.J., Liu, J., Jin, Z.H., Shin, K.S.: Grain boundary segregation and hydrogen-induced fracture in 7050 aluminium alloy. Acta Mater. **44**(8), 3241–3248 (1996). https://doi.org/10.1016/1359-6454(95)00406-8
11. Toda, H., Masuda, S., Batres, R., Kobayashi, M., Aoyama, S., Onodera, M., Furusawa, R., Uesugi, K., Takeuchi, A., Suzuki, Y.: Statistical assessment of fatigue crack initiation from sub-surface hydrogen micropores in high-quality die-cast aluminum. Acta Mater. **59**(12), 4990–4998 (2011). https://doi.org/10.1016/j.actamat.2011.04.049
12. Tiryakioğlu, M.: The effect of hydrogen on pore formation in aluminum alloy castings: Myth versus reality. Metals **10**(3), 368 (2020). https://doi.org/10.3390/met10030368
13. Bhuiyan, M.S., Toda, H., Shimizu, K., Su, H., Uesugi, K., Takeuchi, A., Watanabe, Y.: The role of hydrogen on the local fracture toughness properties of 7XXX aluminum alloys. Metall. Mater. Trans. A **49**(11), 5368–5381 (2018)
14. Birnbaum, H.K., Buckley, C., Zeides, F., Sirois, E., Rozenak, P., Spooner, S., Lin, J.S.: Hydrogen in aluminum. J. Alloys Compd. **253**, 260–264 (1997). https://doi.org/10.1016/S0925-8388(96)02968-4
15. Indeitsev, D.A., Osipova, E.V.: Formation of surface hydrogen layer in pure aluminum. In: Doklady Physical Chemistry, vol. 484, No. 1, pp. 4–7. Pleiades Publishing (2019)
16. Shimizu, K., Toda, H., Fujihara, H., Hirayama, K., Uesugi, K., Takeuchi, A.: Hydrogen partitioning behavior and related hydrogen embrittlement in al-zn-mg alloys. Eng. Fract. Mech. **216**, 106503 (2019). https://doi.org/10.1016/j.engfracmech.2019.106503
17. Bruna, M., Sládek, A.: Hydrogen analysis and effect of filtration on final quality of castings from aluminum alloy AlSi7Mg0.3. Arch. Foundry Eng. **11**(1), 5–10 (2011)
18. Buckley, C.E., Birnbaum, H.K., Lin, J.S., Spooner, S., Bellmann, D., Staron, P., Udovic, T.J., Hollar, E.: Characterization of H defects in the aluminum-hydrogen system using small angle scattering techniques. J. Appl. Crystallogr. **34**(2), 119–129 (2001). https://doi.org/10.1107/S0021889800018239
19. Quirós, C., Mougenot, J., Lombardi, G., Redolfi, M., Brinza, O., Charles, Y., Michau, A., Hassouni, K.: Blister formation and hydrogen retention in aluminum and beryllium: a modeling and experimental approach. Nucl. Mater. Energy **12**, 1178–1183 (2017). https://doi.org/10.1016/j.nme.2016.12.036
20. Yamabe, J., Awane, T., Murakami, Y.: Hydrogen trapped at intermetallic particles in aluminum alloy 6061–T6 exposed to high-pressure hydrogen gas and the reason for high resistance against hydrogen embrittlement. Int. J. Hydrogen Energy **42**(38), 24560–24568 (2017). https://doi.org/10.1016/j.ijhydene.2017.08.035
21. Georgiou, E.P., Cevallos, V.P., Van der Donck, T., Drees, D., Meersschaut, J., Panagopoulos, C.N., Celis, J.P.: Effect of cathodic hydrogen charging on the wear behavior of 5754 Al alloy. Wear **390**, 295–301 (2017). https://doi.org/10.1016/j.wear.2017.08.013
22. Lunarska, E., Chernyaeva, O.: Effect of precipitates on hydrogen transport and hydrogen embrittlement of aluminum alloys. Mater. Sci. **40**(3), 399–407 (2004)
23. Felberbaum, M., Landry-Désy, E., Weber, L., Rappaz, M.: Effective hydrogen diffusion coefficient for solidifying aluminum alloys. Acta Mater. **59**(6), 2302–2308 (2011). https://doi.org/10.1016/j.actamat.2010.12.022
24. Anyalebechi, P.N.: Hydrogen diffusion in al-li alloys. Metall. Trans. B **21**(4), 649–655 (1990)
25. Ishikawa, T., McLellan, R.: The diffusivity of hydrogen in aluminum. Acta Metall. **34**(6), 1091–1095 (1986). https://doi.org/10.1016/0001-6160(86)90219-1
26. Nakashima, M., Aratono, Y., Tachikawa, E.: Diffusivity of recoil-injected or thermally-doped tritium in aluminum. J. Nucl. Mater. **98**(1), 27–34 (1981). https://doi.org/10.1016/0022-3115(81)90384-6

27. Scully, J.R., Young Jr, G.A., Smith, S.W.: Hydrogen solubility, diffusion and trapping in high purity aluminum and selected Al-base alloys. In: Starke, Jr., E.A., Sanders, Jr., T.H., Cassada, W.A. (eds.) Materials Science Forum, vol. 331, pp. 1583–1600. Trans Tech Publications Ltd (2000)
28. Buckley, C.E., Birnbaum, H.K.: Characterization of the charging techniques used to introduce hydrogen in aluminum. J. Alloys Compd. **330**, 649–653 (2002). https://doi.org/10.1016/S0925-8388(01)01496-7
29. Jiang, G.R., Li, Y.X., Yuan, L.: Calculation of hydrogen solubility in molten alloys. Trans. Nonferrous Metals Soc. China **21**(5), 1130–1135 (2011). https://doi.org/10.1016/S1003-6326(11)60832-7
30. Anyalebechi, P.N.: Attempt to predict hydrogen solubility limits in liquid multicomponent aluminum alloys. Scripta Mater. **34**(4), 513–517 (1996). https://doi.org/10.1016/1359-6462(95)00591-9
31. Anyalebechi, P.N., Talbot, D.E.J., Granger, D.A.: The solubility of hydrogen in solid binary aluminum lithium alloys. Metall. Trans. B **20**(4), 523–533 (1989)
32. Anyalebechi, P.N.: Analysis and thermodynamic prediction of hydrogen solution in solid and liquid multicomponent aluminum alloys. In: Grandfield, J.F., Eskin, D.G. (eds.) Essential Readings in Light Metals, pp. 185–200. Springer, Cham (2016)
33. Anyalebechi, P.N., Talbot, D.E.J., Granger, D.A.: The solubility of hydrogen in liquid binary al-li alloys. Metall. Trans. B **19**(2), 227–232 (1988)
34. Talbot, D.E.J., Anyalebechi, P.N.: Solubility of hydrogen in liquid aluminum. Mater. Sci. Technol. **4**(1), 1–4 (1988)
35. Anyalebechi, P.N.: Analysis of the effects of alloying elements on hydrogen solubility in liquid aluminum alloys. Scripta Metall. Mater. **33**(8), 1209–1216 (1995). https://doi.org/10.1016/0956-716X(95)00373-4
36. Ichimura, M., Katsuta, H., Sasajima, Y., Imabayashi, M.: Hydrogen and deuterium solubility in aluminum with voids. J. Phys. Chem. Solids **49**(10), 1259–1267 (1988). https://doi.org/10.1016/0022-3697(88)90184-9
37. Outlaw, R.A., Peterson, D.T., Schmidt, F.A.: Diffusion of hydrogen in pure large grain aluminum. Scripta Metall. **16**(3), 287–292 (1982). https://doi.org/10.1016/0036-9748(82)90354-4
38. Hashimoto, E., Kino, T.: Hydrogen diffusion in aluminum at high temperatures. J. Phys. F Metal Phys. **13**(6), 1157 (1983). https://doi.org/10.1088/0305-4608/13/6/013
39. McLellan, R.B.: Kinetics of hydrogen-monovacancy interactions in metals. Scripta Metall. **17**(3), 417–420 (1983). https://doi.org/10.1016/0036-9748(83)90184-9
40. Smith, S.W., Scully, J.R.: The identification of hydrogen trapping states in an Al-Li-Cu-Zr alloy using thermal desorption spectroscopy. Metall. Mater. Trans. A **31**(1), 179–193 (2000)
41. Furuya, Y., Takasaki, A., Koga, M., Kino, T.: Effect of surface on hydrogen permeation through pure aluminum. In: Materials Science Forum, vol. 331, pp. 1665–1670. Trans Tech Publications Ltd. (2000)
42. Okada, H., Itoh, G., Kanno, M.: Hydrogen segregation in an Al-Li alloy. Scripta Metall. Mater. **26**(1), 69–74 (1992). https://doi.org/10.1016/0956-716X(92)90371-K
43. Tiryakioğlu, M.: Solubility of hydrogen in liquid aluminum: reanalysis of available data. Int. J. Cast Metals Res. **32**(5–6), 315–318 (2019). https://doi.org/10.1080/13640461.2020.1718337
44. Andronov, D.Y., Arseniev, D.G., Polyanskiy, A.M., Polyanskiy, V.A., Yakovlev, Y.A.: Application of multichannel diffusion model to analysis of hydrogen measurements in solid. Int. J. Hydrogen Energy **42**(1), 699–710 (2017). https://doi.org/10.1016/j.ijhydene.2016.10.126
45. Aluminum and aluminum alloys. Methods for determination of hydrogen in solid metal by vacuum heating (in Russian). GOST (State Standard) 21132, pp. 1–98. Izdatelstvo Standartov, Moscow (2000)
46. Konar, J., Banerjee, N.G.: Vacuum heating hydrogen determination in aluminum and aluminum alloys. NML Tech. J. **16**(1–2), 18–19 (1974)
47. Belyaev, A.K., Polyanskiy, A.M., Polyanskiy, V.A., Sommitsch, C., Yakovlev, Y.A.: Multichannel diffusion vs TDS model on example of energy spectra of bound hydrogen in 34CrNiMo6

steel after a typical heat treatment. Int. J. Hydrog. Energy **41**(20), 8627 (2016). https://doi.org/10.1016/j.ijhydene.2016.03.198

48. Polyanskiy, A.M., Polyanskiy, V.A., Yakovlev, Y.A.: Experimental determination of parameters of multichannel hydrogen diffusion in solid probe. Int. J. Hydrog. Energy **39**(30), 17381–17390 (2014). https://doi.org/10.1016/j.ijhydene.2014.07.080
49. Tretyakov, D., Belyaev, A., Polyanskiy, V., Stepanov, A., Yakovlev, Y.: Correlation of acoustoelasticity with hydrogen saturation during destruction. In: E3S Web of Conferences, vol. 121, p. 01016. EDP Sciences (2019)
50. Polyanskiy, V.A., Belyaev, A.K., Yakovlev, Y.A., Polyanskiy, A.M., Tretyakov, D.A.: Influence of the skin effect of plastic deformation on hydrogen accumulation in metals. In: AIP Conference Proceedings, vol. 2053(1), 020011 (2018). https://doi.org/10.1063/1.5084357

The manufacturer's authorised representative in the EU is Springer Nature Customer Service Centre GmbH, Europaplatz 3, 69115 Heidelberg, Germany. If you have any concerns regarding our products, please contact ProductSafety@springernature.com

Printed and bound by CPI Group (UK) Ltd, Croydon, CR0 4YY
30/09/2026
02225289-0003